# The Oath

A Quest for Freedom in War-Torn Ukraine

# The Oath

## A Quest for Freedom in War-Torn Ukraine

**Michael Chemny**

Introduction, Dr. M. Andrew Holowchak

*The Oath: A Quest for Freedom in War-Torn Ukraine*

© *2020* by M. Andrew Holowchak

All rights reserved. No portion of this publication may be reproduced, stored in a re-trieval system, or transmitted by any means—electronic, mechanical, photocopying, re-cording, or any other—except for brief quotations in printed reviews, without the prior written permission of the publisher.

Editors: Liesel Schmidt and Regina Cornell

Cover Design: CC Milford

Interior Design: CC Milford

**Indigo River Publishing**
3 West Garden Street, Ste. 718
Pensacola, FL 32502
www.indigoriverpublishing.com

Ordering Information:

Quantity sales: Special discounts are available on quantity purchases by corporations, associations, and others. For details, contact the publisher at the address above.

Orders by US trade bookstores and wholesalers: Please contact the publisher at the ad-dress above.

Printed in the United States of America

Library of Congress Control Number: 2019948685

ISBN: 978-1-950906-16-1

First Edition

*With Indigo River Publishing, you can always expect great books, strong voices, and meaningful messages.*
*Most importantly, you'll always find . . . words worth reading.*

# CONTENTS

**PREFACE** ............................................................................ xvii

**INTRODUCTION** ............................................................. xxvii

**PART I: MY VILLAGE, KHOM'YAKIVKA** ...................... xxxiii

CHAPTER 1: My Early Years ................................................ 1

CHAPTER 2: Ukrainian Fashion of the Day ........................ 7

CHAPTER 3: My Native Village .......................................... 11

CHAPTER 4: The Christmas Holiday Season .................... 17

CHAPTER 5: The Yordan Holiday Season ......................... 21

CHAPTER 6: The Maslenitsya Holiday Season ................. 23

CHAPTER 7: The Easter Holiday Season ........................... 29

CHAPTER 8: The Green Holiday Season and Lesser Holidays ........... 35

CHAPTER 9: Harvesting the Crops ................................... 39

CHAPTER 10: Mowing after Harvest ................................ 43

**PART II: WORLD WAR I** ................................................. 47

CHAPTER 11: Tragedy Strikes ............................................ 49

CHAPTER 12: World War I Begins .................................... 53

CHAPTER 13: Two Misfortunes, One Day ........................ 61

CHAPTER 14: Third Moscow Invasion ................................................... 65

CHAPTER 15: Revolution in Russia ...................................................... 77

CHAPTER 16: Russian Withdrawal ....................................................... 83

CHAPTER 17: Ukrainian Independence ................................................ 89

CHAPTER 18: Ukrainian Rebirth ........................................................... 95

CHAPTER 19: "I shall faithfully serve my motherland, Ukraine" .... 103

**PART III: POLISH YOKE** ..................................................................... 111

CHAPTER 20: Poles Invade Ukraine ..................................................... 113

CHAPTER 21: "A coffin for me!" ........................................................... 117

CHAPTER 22: A Week in "Dibrova" ...................................................... 125

CHAPTER 23: A Visit by a Polish Commissar ...................................... 129

CHAPTER 24: The Polish-Bolshevik War ............................................. 135

CHAPTER 25: The Black Division Arrives ........................................... 141

CHAPTER 26: Recovery of the Reading Room .................................... 149

CHAPTER 27: The Cooperative .............................................................. 153

CHAPTER 28: Polish Appropriation of Our Lands ............................. 159

CHAPTER 29: "Count Osaslavsky ... wants to see you" ..................... 169

**PART IV: HAVANA, CUBA** ................................................................. 177

CHAPTER 30: "It's hard and painful for me to part with you" ......... 179

CHAPTER 31: To Paris, "Capital of the World" ................................... 183

CHAPTER 32: An Improvisational Ukrainian Concert ...................... 191

CHAPTER 33: Leaving San-Nazaire ......................................................... 195

CHAPTER 34: "I have a visa for Mexico and for Cuba" ..................... 199

CHAPTER 35: *You're here looking for me!"* ......................................... 205

CHAPTER 36: "Somehow, the Lord will help us" ................................ 209

CHAPTER 37: "Lord, please keep us from the storm!" ...................... 215

**PART V: A LAND UNKNOWN** ............................................................. 221

CHAPTER 38: "This Isn't Havana" ........................................................ 223

CHAPTER 39: Settling in Detroit ........................................................... 231

CHAPTER 40: "God is not nonsense!" ................................................... 237

CHAPTER 41: Reunited with My Family ............................................. 243

CHAPTER 42: The Great Depression ................................................... 249

CHAPTER 43: Head Otaman Calls a Meeting .................................... 253

CHAPTER 44: "Like fish on ice" ............................................................. 259

CHAPTER 45: "Business or Politics?" .................................................... 267

**PART VI: TAKING ROOT IN THE USA** ............................................. 273

CHAPTER 46: First Ukrainian Congress ............................................. 275

CHAPTER 47: "You feel a Holy Duty, and you act on it!" ................. 283

CHAPTER 48: World War II and the Demise of the UHO .............. 287

CHAPTER 49: Two Life-Threatening Surgeries .................................. 289

CHAPTER 50: Trip to California ............................................................ 293

CHAPTER 51: The Hetmanich Pays Me a Visit ................................. 297

CHAPTER 52: Another Trip to California ............................................ 301

CHAPTER 53: Death of Danilo Skoropadsky ..................................... 305

**EPILOGUE**.................................................................................................. 309

# ILLUSTRATIONS

1.1 Village Sign

1.2 My Parents, Kornylo and Anna Chemny

2.1 Girls of Komy'akivka

3.1 Church of St. Michael the Archangel, Chemny's Day

6.1 Some Ukrainian Staples

7.1 Church of St. Michael the Archangel, Today

8.1 Church of Mary, Mother of God, Today

9.1 Typical Ukrainian Wheatfield

10.1 Reaping the Harvest

12.1 Russian Soldiers of WW I

15.1 Russian Revolution of 1917

17.1 Ukraine's Push for Independence

20.1 Chemny as Otaman Sich

21.1 Chemny's House, Today

21.2 Efrosina Chemny

24.1 Polish Light Cavalry

27.1 Wedding Picture

31.1 Warsaw Railway Station

32.1 Ship Leaving San-Nazaire

34.1 Havana, Cuba

36.1 Boat at Sea

37.1 A Florida Swamp

39.1 Immaculate Conception Church

41.1 Packard Motor Company

42.1 The Great Depression

44.1 Chemny and Colonel Alexander Shapolval

45.1 The Airplane "Ukraina"

45.2 Chemny on Roof of His House

47.1 Bishop Constantine Bohachevsky with Priests

49.1 Chemny's Family

50.1 Vasyl Yemetz, Virtuoso of Ukraine

51.1 Hetman Toasts Ukraine

52.1 Second Visit to California

53.1 Gravestone of Chemny's Parents

# MICHAEL CHEMNY TIMELINE

30 Oct. 1899: Michael Chemny is born.

1910: Chemny's maternal grandfather dies.

14 Sept. 1914: War begins, and Russian Cossacks ransack Tysmenytsya and Khom'yakivka.

Winter 1914–1915: Russian troops leave Khom'yakivka and Austrians appear for two weeks.

July 1915: Khom'yakivka suffers conflagration.

Spring 1916: Moscuvites return to Khom'yakivka.

Autumn 1916: Chemny forced to be guide for several months to Russian soldiers. When he returns home, his town and house are overrun by Russians. Chemny translates into Ukrainian *The Siege of the Mill*.

Feb. 1917: Chemny receives news of the Russian Revolution.

17 Mar. 1917: Ukraine's Central Rada is formed.

29 Apr. 1917: With Ukrainian uprising, Pavlo Skorpadsky is appointed Hetman.

1 Nov. 1918: Chemny ordered to Pshenychnyky, Nadorozhna, Klubovets, and Ostrynya as a courier to report collapse of Austria and ascension of Ukrainian National Rada.

2 Nov. 1918: Chemny travels with 120 soldiers to Lyads'ke-Shlyakhotskoye and gives speech for Ukrainian unity and independence

3 Nov. 1918: Chemny with some 70 other of his villagers takes an oath to serve Ukraine. He later travels to Stanislav to became part of the First Striletsk Regiment.

## The Oath

21 Nov. 1918: Polish troops drive Ukrainian troops from L'viv and remove to Lynnyky, where Chemny is wounded by a grenade. He is hospitalized till February, 1919.

c. 25 Feb. 1919: Chemny home from the hospital to convalesce.

25 May 1919: Poles advance to Stanislav; Ukrainians desert the village.

June 1919: Poles occupy Khom'yakivka.

8 Feb. 1920: Chemny gets typhoid fever.

19 Apr. 1920: Chemny arrested, beaten by Polish soldiers, and jailed.

May 1920: Polish commissar visits Kornylo Chemny and blackmails him into political service.

Mid-Aug. 1820: Ukrainian Black Division enters Khom-yakivka and causes chaos.

7 Nov. 1920: Chemny effects restoration of Provita (Ukrainian Reading Room).

16 Nov. 1920: Chemny organizes and heads a Ukrainian cooperate—the Zluka Cooperative.

7 Nov. 1922: Chemny weds Theophilia Tymnchyshyn.

22 Dec. 1922: Chemny forced to join Polish Army.

15 Mar. 1923: Chemny, with furuncles, is given a two-year leave from the Polish Army. Goes to Tysmenytsya, his new home, where he arrives on March 21, 1923.

28 Jan. 1924: Chemny's daughter Eustaphie is born.

15 Feb. 1924: Chemny refuses a high Polish post.

16 Apr. 1924: Chemny leaves Tysmenytsya for the New World with companion Demyan.

# Michael Chemny

10 May 1924: Chemny and Demyan arrive in Havana, Cuba.

14 June 1924: Chemny, Demyan, and Fritz survive a raging storm at sea, jump ship, and find themselves in Florida.

19 June 1924: Father-in-law comes for Chemny and Demyan.

Late-summer 1924: Chemny begins work for Detroit Edison for 30 cents/hour. That lasts for 10 weeks. He joins Immaculate Conception Ukrainian Catholic Church and becomes active member of Sich.

Late-Autumn–Winter 1924–1925: Chemny begins work in a pipe factory for 40 cents/hour. Studies English at school and earns diploma. Tears open hand during when pipe splits and is incapacitated.

Spring 1925: Chemny begins work as a spring factory. He begins to contribute essays to Sich. He suffers severe burn on right hand and arm, which incapacitates him for six weeks.

Summer 1925: Chemny begins 11-hour shifts at night at Packard Motor Company, while he at-tends a technical university.

25 Nov. 1925: Mother-in-law and daughters Hanna and Eugenia arrive in Detroit from Ukraine. Packard soon eliminates night shift and Chemny, unemployed, withdraws from school.

Late 1925: Chemny moves to Cleveland to work with his uncle at the latter's factory.

Early 1926: Chemny matriculates at Lincoln University.

3 Apr. 1926: Chemny, after rift with uncle, returns to Detroit.

Spring 1926: Chemny again gets job, afternoon shift, with Packard Motor Company. He begins courses with Society of Polish Engineers which he continues for two years. He joins Michigan National Guard.

# The Oath

Autumn 1927: Chemny begins work for Ford Motor Company, Rouge Plant (morning shift), in the Give Room. He soon earns $1.05/hour and continues study of engineering. Quickly moved to afternoon shift and then to the night shift, so he discontinues studies on engineer-ing. He develops severe stomach and throat pains. He continues in spare time to write for Sich, even under penalty of deportation. He also works with Canadian Ivan Skaletsky to bring wife and daughter to Michigan, and soon rents a house.

26 Sept. 1927: Theophilia and daughter arrive in Detroit.

Dec. 1918: Chemny gets pneumonia and is bedridden for four weeks. Loses job at Ford, and be-gins 11-hour night shift at Baker Company.

4 Feb. 1929: Natalka Chemny is born.

Spring 1929: Chemny buys his first house.

Apr. 1929: Chief engineer of Baker Company is killed in plane crash, and U.S. government re-calls its order for Baker engines and 2,000 people are laid off. The Great Depression begins.

11 Oct. 1930: Letter from Chief Otaman Sich, Dr. Stephan Hrynevetsky and bids Chemny to come to Chicago. The next day, there is a fiery meeting and Hrynevetsky is ousted.

20 Oct. 1930: Chemny asked to teach Ukrainian and Ukrainian culture to the children of Brotherhood of St. Michael the Archangel.

Spring 1931: Michigan aims to purge state of illegal immigrants. Chemny quits teaching at end of May to move to Ohio. Supreme Court overturns decree, but Chemny with family still moves to Ohio.

July 1931: Chemny sells insurance in Cleveland.

Sept. 1931: Chemny's family removes to Detroit and family suffers an automobile accident on the trip.

## Michael Chemny

Dec. 1931: Chemny, after having sold all his furniture, removes to join his family in his house in Detroit.

Christmas 1931: Penurious Chemny receives a basket of food from his church. Chemny continues to be out of work till summer of 1932.

Autumn 1932: Chemny is bidden to travel eastern and mid-eastern states and revitalize Sich interest.

Feb. and Mar. 1933: Writes "Mother and Strilets" and "For Freedom" for Sich.

Early Mar. 1933: Father Kornylo Chemny dies at 60 years of age. Chemny, unemployed, takes to begging on the street for food and money for his family.

Aug. 1933: Chemny begins job as a milkman.

Feb. 1936: Chemny appointed Otaman of District Four of Hetman Sich.

21 May 1937: Chemny begins job at Snyder Tool and Engineering.

May 1937: Chemny begins working for Ukrainian Congress Committee of America and the Ukrainian American Relief Committee.

9 Aug. 1943: Chemny becomes a U.S. citizen.

4 July 1943: Bishop Constantine Bohachevsky visits Chemny.

Nov. 1948: Beginning of Ukrainian Hetman Organization, with Chemny as secretary.

8 Oct. 1949: Chemny has surgery for ulcer and has 7/8ths of his stomach removed. He is bedridden till February, 1950.

Oct. 1949: Chemny is elected vice-president of Providence, the Union of Ukrainian Catholics.

July 1953: Chemny goes to California to visit cousin Olga Miller.

23 Oct. 1953: The Ukrainian Hetmanich visits Chemny.

23 Feb. 1957: Hetmanich Skoropadsky is killed.

24 Dec. 1962: Chemny suffers first heart attack.

18 May 1963: Chemny suffers a slight stroke.

Spring 1965: Chemny ends his biography.

1966: Chemny retires from Snyder Tool and Engineering Company.

Early 1970s: Chemny moves to Redford Township.

12 Nov. 1972: Chemny and Theophilia celebrate 50 years of marriage.

15 Feb. 1979: Chemny dies.

# PREFACE

M. Andrew Holowchak, Ph.D.

ON THE TWENTY-SECOND of June 2017, my mother, Natalie, died. The event was significant in my life for several reasons—one of which was unanticipated. My cousin Steve Fedak brought with him to her wake numerous copies of a partial translation of my maternal grandfather's book, titled Потоптані Мрії—"Trampled Dreams," as it has been customarily translated.

Once I had come back from the funeral, I read through the translation with interest. Finding it fraught with misspellings and numerous incoherencies, I quickly realized that it was horribly inadequate, and so I took it upon myself to clean up what was done of the translation, though I cannot make any claims to large competency with the Ukrainian language.

This cleanup project came to grip me in a way that I could not have anticipated, both because the story was gripping and because it happened to be my grandfather's story. In the span of one week, I came up with a transcription of the translated portion of the book—some forty-five pages' worth of work and the first fourteen chapters. The project necessitated taking certain liberties with the translation, which in most instances reflected problems with the Ukrainian text, not with the translation. And so, deep into the project, I decided to add notes in cases where I had taken liberties with the text. In short, it became for me a scholarly project, not just a pastime.

Fedak and I realized that the book was too important not to be fully translated. Therefore, he sought out and found someone sufficiently acquainted with both Ukrainian and English to offer a reasonable translation of the remainder of the book. I asked for a "literal interpretation" from the Ukrainian in order to preserve ambiguities rather than simply an interpretation.

Once I finished the transcription and recrafting of the translation, it occurred to me that the original title was ambiguous. *Potoptani* (потоптані) typically means "trampled," but it can also mean "deceptious" or "deceitful," in the sense of something appearing to be greater than it is. It is important that readers are apprised of this slight ambiguity of the term as they read the book, as it might be that Chemny had in mind "Deceitful Dreams" and not "Trampled Dreams." The difference between the two is slight, but the choice of translating one way or the other is not without implications. It might also be the case that he chose the word because of its ambiguity.

After having read through, transcribed, and smoothed out the text, one thing immediately became obvious. My grandfather, who always regarded himself as a writer—he recalls an episode in chapter 14 with Colonel Dudinsky, who praised Chemny's translation of a small book called *The Siege of the Mill* and encouraged the young man to continue writing—was far from a polished writer. The book, overall, had little structure. There were numerous divagations, and events were not related with an eye to chronological consistency or logical smoothness.

First, Chemny often disregards minutiae when they are sorely needed. Descriptions of key events are sometimes explained in detail, but most are often given short shrift. I offer some examples.

When Chemny writes of his childhood, he oddly fails to mention how many siblings he had. Only by a thorough reading of the book does one find that he had two brothers and a sister. This is consistent with Fedak's genealogical research, which shows that Chemny had three younger siblings—two brothers, Volodymyr and Yaroslav, and one sister, Efrosina or Rusia. Chemny mentions Volodymyr by name five times (chapters 16 and 23, 24 [twice], and 25) and only once does he mention Efrosina (chapter 21). Yaroslav is mentioned by name only in chapter 23. In the overwhelming number of cases, Chemny merely uses "brother" and "sister." Did he dislike them, was he disliked by them, or did he merely wish to keep the focus on himself? The same can be said of his mother, who is relatively silent through the book, except when Chemny takes ill with typhus in chapter 20 and when his mother finds him a wife (chapter 29).

Moreover, there is never a description of his house or the layout of his farm. It is clear, as one reads through the text, that his family was better off than most others. His father was literate, had a library, and wanted his son to be educated, and Chemny, as a boy, was certainly educated to some degree. The war intervened in 1914. He was supposed to go to Vienna to study, but World War I came along and he stayed on his farm. Given that they were better off than most other villagers, the house must have at least been larger and somewhat more elaborate than most others. It is the same with the farm. It is unusual that the house and farm, which are referred to so frequently in the text, are never described in any detail.

Furthermore—and this is the most egregious omission in an autobiography—he devotes only one paragraph to his marriage, which was prearranged by his mother. He writes: "She [my mother] even found a girl for me: Miss Theophilia Tymchyshyn. I liked Theophilia. She was not only hardworking but also beautiful. We were introduced and soon fell in love. On November 7, 1922, we got married." That is all. One would expect in an autobiography much more text—perhaps even a chapter—devoted to her, given the enormousness and significance of weddings in Ukrainian culture and given the centrality of family. This does not happen. This is not to say that he does not write of her thereafter with tenderness and great affection. He does. It is merely that, as he tells his story, she does not seem to be a part of the events of which he writes, and we know that that was not the case.

Second, the account of his life is *Chemny's* account. Chemny writes with jaundiced eyes, and that tells us much about Chemny's *Weltanschauung*. We do not find people in the story, but good and bad people. The good ones are God-loving and God-fearing traditionalists and Ukrainian nationalists; the bad ones shun religion and tradition and are indifferent to the fate of Ukraine. Again, I offer some illustrations.

The village priest, Father Durdello, has almost godlike status for Chemny. Chemny has no harsh words for the priest even when he packs his belongings, gathers his family, and flees the village in fear of persecution by Russian soldiers at the advent of World War I. Chemny writes with disgust at the pillaging of the priest's farm by his fellow villagers after the priest flees Khom'yakivka. Here some detail is not wanting. When the priest, for whatever reason, returns in three days, young Chemny, alone of all the villagers and who is but fifteen at the time, courageously enters the small chapel—the one coincidently built by Chemny's father—where the despondent priest has gone to pray, and Chemny also prays in the chapel, behind the priest. When the priest comes to the young man and expresses his sorrow concerning the pillaging, young Chemny has sagacious, Christ-like advice for the priest: "Father, forgive them, for they know not what they are doing." Chemny never considers that at least part of the reason for the villagers' plundering the priest's farm after he had left is revenge for him, their spiritual leader, having abandoned the villagers in an especial time of need.

Early on Chemny writes contemptuously of Jews, presumably because they do not embrace the divinity of Christ. During the Christmas holiday, a young man is selected to wear a Jewish mask, dress like a Jew, and parade around the village, while Chemny and his friends tell crass jokes. On Bright Tuesday after Easter, some boy would again dress as a Jew and chase after, hug, and even kiss—presumably through his mask—a Ukrainian maiden. The girl would then spit on the lad in contempt and slap the boy. The ritual, Chemny acknowledges, is "weird," but "very funny." Yet later in the book, when Chemny trudges through Warsaw in pursuit of something to eat, two Jews give him bread, herring, and some pocket change, and he acknowledges that there is more of the Good Samaritan in them than in all the Christian Poles in Warsaw.

# Michael Chemny

Chemny also writes paternalistically, perhaps contemptuously, of his fellow villagers by using the word *peasants* (селяни[1]) as if to distance himself from them. He says that the villagers called him "rich boy," and he struggled to understand why he was given that name, as he considered himself much like they were. Yet we come to find that his family had more than most in the village. At Easter, his mother forced him to go out and share butter and cheese with those neighbors less fortunate than them. Moreover, he, unlike most, was literate, and his father does not seem to have had to work the field with the laborers. He was instead a builder (chapter 11). For instance, he describes his father watching the beautiful machine-like efficiency of his laborers during harvest. When Chemny escapes from Polish yoke and takes as a travel companion youthful Demyan from his village, he introduces him as "unkempt and arrogant," and says little about the teenager throughout their trip from Ukraine to France to Cuba and to Florida. They part on the ride up from Florida—Demyan to Cleveland and Chemny to Detroit—and we later hear nothing of Demyan. Did they correspond thereafter?

Moreover, there is only one account of young Chemny having fun with the other boys of the village—he never mentions a best friend or any friends—and so we recognize that fun was just not much a part of his life. After the death of his grandfather, to whom we are introduced in the first chapter, he, though still a young boy, becomes a sober, serious person. He thought of himself, and very likely was taught to think of himself, as better than the other villagers, and he was likely very much disliked by others, perhaps even his own younger siblings.

As a sober and serious person, he writes soberly and seriously about significant events in his life. There is one puzzling exception. In chapter 51, the second from the last chapter, he describes a second trip to California—a pleasure trip to Malibu Beach with his dear friend, the Ukrainian virtuoso Vasyl Yemetz. Chemny spends almost the entire chapter describing a bad sunburn he gets while lying on the beach. "By evening, my leg, which had been scorched under the sun's rays, was nearly beet red and I felt such a burning pain—a pain of such intensity that I had never before felt—that I could not sleep." That comes from a man who suffered a scorching burn in the pipe factory and a mild heart attack and had seven-eighths of his stomach removed. Why does this event deserve such especial attention? The scenario at the beach is made almost comical by the fact that only his left leg was burned—from the top of his leg to the foot. How one could expose only one leg to the sun is something he never explains. Was he asleep? Was he drunk?

In some respects, Chemny is to be pardoned for his biases. The book does not read like an autobiography, and it is perhaps best not to take the book as an autobiography—at least, not exclusively as one. The early chapters on Ukrainian dress and life in Khom'yakivka when he was a boy are not in the least about Michael Chemny. They are written, sometimes with remarkable clarity, to capture for posterity life in his village in the early twentieth century, prior to the travesty of World War I.

When I first went through *Trampled Dreams,* I ordered the various chapters into two main parts, titled "My Life in Europe" and "My Life in America," as the book is, in effect, the story of a life in two different worlds, with his dreams in each trampled. Yet, as I thought more on that partitioning, it seemed inadequate, as, in essence, the book is about war and its motley effects on people—Chemny especially—and that bipartition seemed not sufficiently to capture that. So I broke the book into six parts so that the centrality of World War I and its effects on him would be more evident.

The first part (chapters 1–11), "My Village, Khom'yakivka," concerns Chemny's account of his childhood in his village in Ukraine. The first three chapters describe his village, pleasant memories of times with his maternal grandfather, and the colorful and ornate Ukrainian dress of his villagers. He then takes us through the agrarian life in villatic Ukraine by ushering us through the various religious holidays, which function as stops to the yearlong drudgery of the life of a village farmer.

Part II (chapters 12–18), "World War I," highlights the chaos in Khom'yakivka and neighboring villages and towns as the war begins. Chemny writes of the Third Moscow Invasion as well as the Russian Revolution and Russian withdrawal. In chapter 18, Chemny proudly and movingly takes an oath as part of membership in the Ukrainian army in some effort to free Ukraine.

The third part (chapters 19–28) is about the Polish invasion of Ukraine, the Polish-Bolshevik War, and its destructive effects on Chemny's life and his own family, as he takes a wife and soon has a daughter. Chemny is shortly forced into the Polish Army and then given leave because of severe health issues before he and his father are coerced to work for the Polish government that occupies Ukraine.

In Part IV (chapters 29–36), Chemny secretly leaves Ukraine with the hopes of somehow making his way to the United States, where he has family in Detroit and Cleveland. As a fugitive from the Polish government, he and a young companion, Demyan, furtively and cautiously make their way to Warsaw, to France, and then to Portugal, where they board a ship to Cuba. Penniless in Cuba, Chemny and Demyan, for two pesos each, board a boat at Havana that is to make its way to Cárdenas so those on the boat can make some money by picking crops. The boat's engine quits, and the thirty or so passengers find themselves stranded at sea for four days with undrinkable water and almost no food.

"A Land Unknown" is the fifth part (chapters 37–44). On the third day at sea, there is a parlous storm, which turns everyone's attention to God. Surviving the storm, Chemny, his young companion, and a German friend named Fritz jump ship early in the morning and make a long and treacherous swim to land. Thinking themselves back in Cuba, they are astonished to find themselves in Florida. Chemny soon finds his way to Detroit; Demyan, to Cleveland. Life in America, a land much unlike Ukraine, begins. Having saved up enough money, he sends for his family, also escaping Polish yoke. His new life in America greatly concerns matters back in Ukraine.

Part VI (chapters 37–44), "Taking Roots in the USA," ends the book. These chapters catalog much of Chemny's involvement in Ukrainian-American politics. Chemny writes of certain highlights of his difficult life (visits by the Ukrainian bishop and by the Ukrainian *hetmanich*), two pleasure-trips to California, and a marked decline of his health, which ends the book. The year is 1965, and readers are led to believe that he would not live much longer. He would, however, live another fourteen years.

The book, despite its defects, is of immense historical value—if only because it offers a snapshot, even if a jaundiced one, of life in villatic Ukraine in the early twentieth century, before, during, and after World War I. I know of no other account of life in Khom'yakivka at the time by any other author. I am proud to have had an important role in preparing this book for publication. It is also of immense historical value because it is a story of one man's quest to find freedom for himself and his family by coming to America—a country where government does not dictate to its citizens how they ought to live. This is a story that millions have lived, but unfortunately a story about which too few have written, hence its value. It is also a story of one man's quest to find freedom for his country, Ukraine, as Chemny tirelessly works while in America for Ukrainian independence and restoration of Ukrainian values.

A few procedural words before closing.

I was aided immeasurably by the accepted transliterations of Ukrainian words, found on the Internet, and thus altered the transcription to accord with commonly accepted transliterations. I used Google Maps for transliterations of Ukrainian towns and villages (e.g., "Homyakivka" became "Khom'yakivka" and "Tysmennitsa" became "Tysmenytsya").

Also, I sought the advice of fellow Ukrainian Dr. Irene Gilliland, at the University of the Incarnate Word, for some grasp of the holiday of Myasnytsy. Here the Internet was unavailing. After some digging, she informed me that this holiday was better known as Masnytsya (Масниця), also known as Kolodiï (Колодій) or Maslyana (Масляна). Given this, everything fell into place and the Internet was availing.

Again, searches on Google revealed mistakes of the translator. For instance, *gorilka* showed itself to be *horilka*—the difference between the *h* sound and *g* sound represented subtly in Ukrainian Cyrillic by a downturn of the horizontal stroke on the *h*-sounding letter and an upturn on the horizontal bar of the *g*-sounding letter (i.e., Гг versus Ґґ). It is the same with Grinevetski and Hrynevetsky. I had thought the two to be distinct persons, until the context showed otherwise and revealed the mistake of the second translator.

Finally, a few words on the liberties I have taken with the text—something that literary purists might find objectionable. My reasons are several.

First, a literal translation would make the book difficult to read. For example, Chemny discusses but glosses over key events like Russia's involvement in World War I and its implications for Ukrainians and the Bolshevik Revolution. So, it was necessary to add something to the discussion to give readers more background. I was helped immensely by the resource the Internet Encyclopedia of Ukraine, as well as several general historical encyclopedias, which enabled me to place events insufficiently described in a fuller context for the benefit of readers. Again, there are certain gaps in his story that needed to be filled so that readers could better follow his account. It was often the case that such gaps could only be filled by working further into the text and then returning to the puzzling lacunae. In the main, where I add material, I give a footnote to tell readers that the material is not Chemny's. Furthermore, the book is not always chronologically consistent. I have made it chronologically consistent, and that required rearranging certain chapters. Finally, Chemny often throws together several themes or events in one chapter, and that makes the given chapter's title dubious. I have often broken such chapters into several smaller chapters, in keeping with Chemny's general tendency to craft short chapters. And so, where sections of a chapter are separated by Chemny in the Ukrainian version thus, ***, I have most often constructed new chapters.

Second, I have added a brief epilogue. Chemny's account ends in 1965, but he lived till 1979. Much happened in that time. His health declined further—there were strokes and heart attacks—and he and his wife and sister-in-law, Anna, moved to a smaller residence on Sioux Street in Redford Township—west of Detroit, where their daughter Natalka, my mother, lived. I have added what I could dig up of that fourteen-year period, though numerous events, some significant, have been lost to posterity.

Finally and most significantly—and here, many in my family might think I have taken too great a liberty—I have changed the title of the book from *Trampled Dreams* to *The Oath*. Why did I change the title? The most significant reason is this: Chemny's story is about triumph through persistency over a lifetime. He saw much and suffered much, but he ultimately prevailed in circumstances that would have felled a lesser man. *Trampled Dreams* implies failure, not triumph. *The Oath* is neutral.

Before closing, I must acknowledge some debts. First, there is the debt to my distant cousin John Renock (Hrenuik) from Ohio, who has been instrumental in finding a literal translator to translate Chemny's book from Ukrainian to English and has helped to track down Ukrainian relatives. Second, there is a debt to my cousin Stephen Fedak, who has been working on our family tree and who threw into my lap, at my mother's wake, a translation of about one-fifth of the book. Last, there has been the aid of my brother David, who has generously given me advice for improvement of my work.

—M. Andrew Holowchak, Ph.D.

# INTRODUCTION

M. Andrew Holowchak, Ph.D.

*THE OATH* IS ABOUT one man's paradise lost—a man robbed of a way of life. Life in Khom'yakivka for Chemny is Elysium. Work is exhausting but rewarding, and there are numerous breaks in the yearly work-cycle through religious holidays. There is a healthy, clean regularity to the year, to daily affairs, and even to religious holidays, which sometimes last three weeks—the length of celebrating a typical wedding. Villatic life is harsh yet simple and, for Chemny, happy. When it is time to work, the villagers work. When it is time to play, they play, and they play with the same intensity, even strenuousness, with which they work. Chemny comes to recognize early on the great beauty of living simply and of comity—he seems to say during times of great misfortune that the aim of life, the beauty of life, is sharing sorrows to lighten their load (a very Epicurean sentiment)—but his fellow villagers never come to grips with that recognition. Chemny adopts a working-man's ethic, founded on the notions of devotion to God, respecting tradition, working hard, living simply and honestly, loving and serving the motherland, and helping others to endure inevitable hardships and misfortunes. I offer some instances.

Chemny begins his book with an account of his shared adventures with his maternal grandfather, Roman, who took him into the woods to pick berries, read stories to him, and played with him. His grandfather passed while he was still a boy, and Chemny paints a vivid picture of the funereal pageantry—the procession, where he and his family walk behind the coffin and the entire village comes to pay their respects, and the gilded coffin, a present of the prince of the region. Chemny writes:

# The Oath

> Like the man, larger than life, his funeral was majestic. I remember it still as if it were now happening. It is still fresh and vivid in my memory. Three priests were reading from the Bible as the grand coffin, framed with gold and a gift from Prince Lyubomirsky, was being pulled by four regal horses. Everyone from Khom'yakivka attended the funeral. My father, mother, other relatives, and I proudly, but sorrowfully, walked behind the coffin. The rest of the townspeople followed.

The funeral is described with such detail—found in few places in the book—that it is certainly a sign of the boy's love for his grandfather and the pain he felt on his passing. How long and frequently he must have cried after the loss. It is likely a loss from which he never recovered—one which catapulted him to manhood much too early.

Chemny throughout is a traditionalist, not a revolutionist or revisionist. He loves Ukraine and Ukrainians' way of simple villatic living, and he is willing to die for his motherland, under Polish oppression. His description of inscription in the Ukrainian army in chapter 18, and how he felt after the oath he swore with his fellow soldiers and before the priests and God, is heartrending.

> I was no longer standing on the ground, but I was somewhere in Heaven and was the happiest of all people in the world. At that moment, I fully experienced my whole being. It was something that I cannot now express in words—perhaps something that cannot be captured by words. I can say that it was the happiest moment in my life. Never before had I experienced such bliss and mental wellbeing, and I cannot expect I shall ever again experience such a blissful moment. With that oath, my world widened—eternity was in a sense captured in a flicker of time—and to this day, I mumble in my heart and head the gist of the oath and shall take that memory to my grave.

# Michael Chemny

One cannot help being envious of a person, if only still a very young man and even if the passage is an expression of great naïveté, who experienced such a moment. I never have. I suspect only few of us have.

Chemny is simpleminded, perhaps even naïvely so. His love of God and religious tradition are unconditional. He finds meaning, even profound beauty, in simple agrarian work. He recognizes the absurdity of war—decent people can be moved to unspeakable, almost discretionary, violence—yet he recognizes its necessity, if only to try to preserve a way of life, uniquely Ukrainian.

Through the early chapters, we become aware of the fragility of life in his villatic Ukraine. World War I forever ends all hopes of happiness.

A man of large passions and of deep courage, his dreams, though simple, could only have been trampled or proven deceitful over time in Ukraine. We feel Chemny's pain as he struggles to come to grips with the atrocities and the twists of fortune during war. We feel Chemny's indignation at the looting done by fellow Ukrainians on their own people in the town of Tysmenytsya at the bidding of Russian Cossacks, but share his sense of pride as he, refusing to be a part of the thievery, walks home from the ugliness. We feel Chemny's sense of beauty when he describes the music of the scythes, "the music of metal," as the Ukrainian laborers reap the oats, rye, and wheat. We feel Chemny's pride as he swears an oath before God to fight for Ukrainian freedom, even at the cost of his life. We feel Chemny's profound shame when he talks about Ukrainian deserters in the war with Poland—some sixty of whom were from his village. We feel Chemny's disappointment that the village priest, Father Durdello, leaves the village for the mountains when he loses his faith in the villagers. We feel Chemny's authenticity as he refuses a post in the Polish government—a post which would have made him a wealthy man—to manage unsettled Ukrainian affairs. It is astonishing that through it all—with everything he suffers—he keeps himself together. That he does so is likely the consequence of daring to have a dream and to follow it at any cost.

# The Oath

Chemny is a boy who is forced by misfortunes to be a man much too soon, but that may have been common among boys in the villages of Ukraine at the time. This is something that most of us today just will not be able to understand. Many of us today have no desire to grow up. We wish to extend our childhood as far as possible into adulthood. We work, but find no meaning in work, and work only to have time to play. We want the thrill, excitement, and kick of jumping from an airplane, owning and driving a sporty vehicle, or climbing a mountain. The world in which we live is just too different. We could never hear, as did Chemny, music in the scything of rye.

The Polish occupation of Ukraine and numberless Polish atrocities force Chemny to leave his beloved Ukraine in search of a new paradise. Yet the New World proves radically different, though the story is much the same: His dreams in America will never be fully realized. They cannot be, for America is not Ukraine, and Detroit is not Khom'yakivka. Chemny settles in Detroit, but has difficulty finding work. When he does find work, it generally involves lengthy shifts, often requires working the afternoon or night shift, unanticipated changes of shifts, and insufficient pay, and such things make impossible his dream of finishing school and becoming an engineer.

Then there are the many lean years—e.g., the Great Depression—in which Chemny often goes days with little to eat. At one point, he moves in with his uncle in Cleveland to take a job selling insurance in some effort to eke out a living for himself and his family. The job pays poorly—no one is going to buy insurance when he is having difficulty putting food on his plate—and Chemny has to send his wife and children back to Detroit, where they can be with her parents, so that they can have food. He manages by selling all their furniture, but even that does not go swimmingly, as many purchasers agree to buy only if they can pay in installments. Unable to get along with his uncle—an atheist with communist sympathies—he, too, eventually moves back to Detroit on Easter weekend.

## Michael Chemny

Enduring the lean years, Chemny works various jobs—laborer at Packard Motor Company, laborer at Ford Motor Company, teacher of Ukrainian at his church, and milkman—until he is hired by Snyder Tool and Engineering Company, an ideal job, where he works dutifully until his retirement.

All the while, Chemny is much involved in Ukrainian political and religious affairs—the two are inseparable for him, and the political groups with which he works are religiously grounded. He works for Sich, Motherland, Bulava, the Ukrainian Congress Committee of America, the Native School, and the United Hetman Organization of America. In all such groups, as he says, "I was always picked for high posts." His work does not go unrecognized. On July 4, 1943, Reverend Bishop Constantine Bohachevsky comes to Detroit and pays Chemny a visit. On October 22, 1953, the hetmanich, Danylo Skoropadsky, comes to Detroit and pays Chemny a visit. Each proves to be a singular honor.

Those singular honors notwithstanding, the freedom he eventually finds in America, once he is a citizen, brings no happiness, because it can bring no happiness. Deracinated in Ukraine by war-linked atrocities, he cannot take root in America, because he is a plant that requires Ukrainian soil—hence Chemny's choice of *Trampled Dreams* as his title.

I close by returning to a point I make in the preface, where I note two flaws of the autobiographical text: disregard of minutiae and Chemny's jaundiced eye, which tends to polarize people and events. Those are mitigated once we realize that the book is more of a philosophical text than it is an autobiography. The theme throughout, as I have noted, is that of a paradisiacal manner of living, described fully in chapters 2 through 10, that is lost because of human greed, lust, desire, and, worst of all, human ignorance—all enflamed by events prior to, during, and after the Great War. For Chemny, those human foibles are seemingly part of the nature of humans, as are their opposites: generosity, self-control, satiation, and knowledge. For Chemny, each man has to fight a war within himself, and

# The Oath

those who win that war, the good men, have to fight those who lose that war, the bad men. Thus, war between peoples is inevitable so that there is some good in the world.

# Part 1
# MY VILLAGE, KHOM'YAKIVKA

# Chapter 1

## My Early Years

Picture 1.1. Village Sign
(photo courtesy Stephen Fedak)

# The Oath

JUST BEHIND THE SLOPES of the Podilyska Highlands, the Pokutska Plain begins. It starts in the town of Tysmenytsya, through which the Vorona River—the branch of the Bystrytsya-Nadvirnyans'ka River to the east—runs. Both of the rivers run into the Dnister River and toward the Carpathian Mountains, which look like a distant blue belt on a good, clear day.

About five kilometers south of Tysmenytsya and above the meandering Strymba Stream, there is a little village, Khom'yakivka, of about two hundred spread-out farms. The village sits by the much-trodden road, Tysmenytsya-Kolomyya (now, H-10). It was founded at the end of the thirteenth century by a certain Khomyak (Хом'яком), whose family lineage died long ago. My great-grandparents settled there after the Yosyfynska Reform.

My grandfather, as he often told us, came to Khom'yakivka from the village of Budzyn, which was southeast of the winding Dnister River.[2] At Budzyn, he had a farm called Budzynovsky (Бузиновсь). Yet when his father lost his farm as well as his noble house, my grandfather began to call himself Paul Chemny.

My grandfather started to work as foreman of a woolen-cloth factory, the owner of which was Prince Lyubomyrsky of Dubno, more than two hundred kilometers north of Khom'yakivka. He married a young widow named Yustina Zubaly, and they bought a little piece of land. Thus, they began their life in Khom'yakivka, where they had five daughters and two sons: my father, Kornylo, and his brother, Michael.

Kornylo married Anna Hryniukivna, the daughter of Roman Hryniuk, who managed all of the woodlands of Prince Lyubomyrsky as well as his own farm. They birthed me—the first of four children. My birth certificate was signed by Father Verhanovsky[3] on October 30, 1901. My parents, however, confirmed that I was born two years earlier. Father Verhanovsky did not like to enter immediately the names of newborn children to the certificate-book—he merely took little notes, which were pretty often lost. Father Michael Durdello corrected the certificate-book in 1902. It was not until I went to high school that I found out that I was two years older than I had thought I was. Since I was somewhat short, the discovery did not bother me.

**Picture 1.2. My Parents, Kornylo and Anna Chemny**

My maternal grandfather, Roman, used to take me to the woods during the summer when I was a child. I used to dart through the bushes all day and pick wild berries, mostly strawberries. When I was tired, I would sit somewhere and lower my head. Roman would pick me up and lay me down me on a bed of fragrant hay, where I had a very sound and deep sleep.[4] When evening came, he would wake me and say tenderly, "Mickey, get up. It's time to go home." Still feeling sleepy, I would rub my eyes and struggle to rise from the bed of straw. Once I was up, my grandfather and I would walk to Khom'yakivka, where he returned me to my mother. He would tell her to give me something to eat and then put me to bed. He would add, with a smile, "He had a good day of play."

I visited the woods with my grandfather very often. I loved both—the woods and my maternal grandfather, Roman—with the whole of my youthful soul. Grandfather taught me about the various trees, bushes, flowers, and even the grasses. He made sure I knew about those four types of plants. I remember vividly him teaching me that the trees, especially the larger ones on the north side, were covered with moss. "Remember," he used to say, "we are now west of the town, and to go back into town, we have to travel in an easterly direction. So the moss will always be on your left side."

"All right, Grandpa," I used to answer with a smile.

I seldom roamed far from Grandpa in the woods, but sometimes, while chasing a rabbit or a beautiful bird, I would roam too far. "Mickey, where are you?" my grandpa would cry.

I usually answered immediately, and thus, we would quickly find each other.

Thus, I would pass the summers quickly, delightfully, and without a care with Grandpa in the woods.

I soon began to go to school, and I missed Grandpa. During summer vacation, I would go to him. With a book under my arm, we would go into the forest, where we would sit wherever we could find a cool place. There, I would read to him different stories from my schoolbook.

On the hot days of July, we would go to the water mill, which was not that far, where I loved to swim. Grandpa would carefully watch me so that I would not go into deep waters.

Oh, how delightful were those days!

Yet when I turned ten years of age, my grandfather caught a cold, got very ill, and died. The year was 1909.

Like the man, larger than life, his funeral was majestic. I remember it still as if it were now happening. It is still fresh and vivid in my memory. Three priests were reading from the Bible as the grand coffin, framed with gold and a gift from Prince Lyubomyrsky, was being pulled by four regal horses. Everyone from Khom'yakivka attended the funeral. My father, mother, other relatives, and I proudly, but sorrowfully, walked behind the coffin. The rest of the townspeople followed.

I was devastated by the death of my grandfather. How often and long I cried. To make matters worse, he died late in spring, when I was awaiting another wonderful summer in the woods with him with excitement. That summer, when I was away from school, I went by myself into the woods and imagined, as a way to console myself over such a great loss, that he was with me and that we were spending together a few more days in the woods.

In the winter of the same year, I also lost my paternal grandfather. His death was not to me such a large loss, perhaps because he had been seriously ill for about two years. Still, he loved me much and would often tell me various charming stories, to which I always listened with full attention.

After Grandpa Roman's death, I became a serious student. I passed an entrance exam to the local high school, where I was accepted without much of a fuss. Studying was easy for me, perhaps as a diversion from the loss of my grandpa.[5] The professors liked me, and I liked the school and the general milieu.

# Chapter 2

## Ukrainian Fashion of the Day

**Picture 2.1. Girls of Khom'yakivka**
Included are Chemny's future wife,
Theophilia (center), and her sister, Anna (right).

# The Oath

O N SUNDAYS AND HOLIDAYS, The village farmers[6] (селяни) dressed neatly and finely insofar as they could afford to do so. There was a sort of competition between many to see who could dress most smartly—smart-dressing being a signal of status.

Still, all cherished proudly and loved Ukrainian traditions and customs. Girls and young women wore the traditional embroidered long shirts, which covered lavish and rich skirts of rose, black, blue, and even green. Beneath their skirts, they wore underskirts of lighter white-colored fabric called the crown (короикою). Each wore black boots. Years prior to World War I, bright red and yellow boots were not uncommon.

In cold weather, village women wore black common pants, without decorations. In winter, they added beautifully decorated long coats with tall collars or shorter coats (киптарями) with long sleeves. They also placed rough wool shawls on their head. Those shawls were mostly colored red and raspberry pink, but other colors were common. Older women distinguished themselves with darker shawls—black was common—but even such drab colors were vivified by bright flowers, weaved into the fabric.

In summer, girls placed nothing on their head. They merely wore their hair in a handsome braid, with bright-colored flowers. To the braid, there were beautiful ribbons that tangled from their head. Little tufts of hair were often curled on their forehead.

Older men wore wildly long shirts, rarely embroidered. They also preferred large, wide black boots and a protective black vest or small coat, depending on the coldness. In summertime, they wore large hats made of straw. In wintertime, they wore tall, black leather headings with fur sewed to the inside.

Younger men wore wildly long down-to-their-knees shirts, which covered their pants. These shirts were exquisitely embroidered, especially on the collars and cuffs, and elaborateness of the embroidery was a sign of wealth. On their waists, younger men used a belt in such a manner that the shirt clung tightly on the front and loosely on the back. They often wore a wide red ribbon under the shirt's collar, which hanged down freely. They, too, wore large black boots. Concerning headwear, they wore a flower-decorated hat in the summertime and a tall black heading with fur. For the most part, young men, as it was the custom and a sign of hardihood, shunned coats, unless the cold was insufferable. When the weather was warmer, young men often merely wore the rough black protective vest, worn by all on colder days. A youth wearing only a black vest stuck out from other Ukrainians, but that was the custom.

All the clothing—besides the scarfs, shawls, boots, and coats—was handmade. The exceptions were bought in Tysmenytsya, which was renowned for its shoemakers and merchants. The villagers could get the same sort of clothing from a neighboring town about two miles to the southwest, Markivtsi, only the shirts were longer. None of the other neighboring villages and towns had such beautiful and elaborate clothing—especially, embroidered shirts.

My family dressed exquisitely, and so we distinguished ourselves from residents of other towns and villages. People said of us that we dressed like the people in big cities.

While still a small boy, I wanted merely to dress like the other village boys and not to stand out from them. I much enjoyed playing with them in the wild fields, yet they called me "rich boy," though I thought I was more like them than different from them.

Still, I never went into the village as did other boys and their older neighbors, who would always come back from the village with fascinating stories of the strange things that they did when all the villagers, tired from work, slept.

# Chapter 3

## My Native Village

Picture 3.1. Church of St. Michael
the Archangel (Chemny's Day)
The boy in the picture is a youthful Chemny.

# The Oath

IT IS NOT SIMPLE to describe our Ukrainian village. A short description fails to capture its pulse. A lengthy description is fulsome, but it, too, can fail to capture its pulse by drowning readers in details. Thus, I prefer a middle path. I offer a brief description of some things, which readers might find not so important, and a more elaborate description of others, which they will probably find more important. In such a manner, I aim to give a full image or picture, as it were, of village life in Khom'yakivka, because that Ukrainian village is so intriguing, multidimensional, and, to my mind, incredible.

Yet where am I to begin? I think it is best to proceed chronologically and look at events over a particular span of time. I begin with late autumn, just before Christmas, in 1911, when I was twelve, and end with early spring.

Before Christmas Lent, all the villagers fasted—called St. Philip's Fast (Пилипівкою)—though they had much work to do in and around their houses and farms. They threshed the crops to get the grain, tied together the straw (околоти) into a wisp (пуки) from which they made sheaves to be used for covering the roofs of their houses. Most people used rye straw, as it was the most durable covering for a roof.

When they were done with the threshing, the villagers began stocking up fuel for the winter and laying fodder for their cattle. They readied potatoes, beets, and threshed oats.

Women, young and older, spun yarn and embroidered shirts for themselves, for their fiancés, and for relatives for the holidays. They gathered in the largest and best-illuminated house and worked while singing folk, church, and love songs. Those so-called parties were sometimes visited by young men, who used to come to sing or joke. Other young men came to look for distinct pieces of straw for new hats for Easter holiday. Still others brought books with attention-grabbing stories or fables and read them aloud. Though the young men visited, the evenings were always civil, non-obstreperous, and scandal-free, and that is because those who owned the houses at which the parties were held managed well the evening.

Many of the youths and young farmers visited the small library, Prosvita, where there were often useful and stimulating readings on farming, history, and politics. These meetings were headed by Father Michael Durdello and my father. Besides the library, there were a general school and Ukrainian school, and the town store. All were managed by the village committee, chosen each year by vote.

There was a new and radical organization in the village—the Sich—which was rapidly growing. Members of Sich (Січ) were young men who worked as firemen. Holding ribbons, a flag, and hammers, they marched to church to attend service every Sunday. Their presence, however, was disruptive, as many were godless, or nearly so. Father Michael Durdello, recognizing that they were challenging the authority of the church, often bid them to stop. He saw them harming both themselves and others. Yet Sichers (Січи) ignored the priest, and even poked fun at him. They audaciously claimed that they would do just fine without a priest and a church. Other churchgoers saw their defiance as a challenge to God. On one occasion, Father Durdello stopped Sunday service and went with cross in hand outside the church to challenge Sichers either to attend mass and conform to traditional customs or to remain permanently away from the church. The Sichers immediately dispersed. Each feared that Father Durdello would recognize him and affect some sort of punishment. Eventually, however, they stopped attending Sunday service.

There was a tavern, owned by a Jew named Aba Brown in the village. It was located at the crossroads of Khom'yakivka. As the Sichers stopped going to church, many turned to the tavern. Yet Father Durdello fought inebriation through formation of the Sobriety Society in Khom'yakivka. That had some influence in Sichers, as some left the radical group.

# The Oath

Father Durdello was an artful speaker who could, if he wished, punish individuals with the right words, and he did just that to Sichers and those frequenting the tavern. He and Aba Brown thus became enemies.

With the growth of the village and groups such as Sich, the villagers become less spiritual and more secular. The village was visited by radical speakers such as Dr. Bachynsky. Makooh, my father's friend, stayed often in our house, especially during elections to the senate or to the Austrian Parliament. He, too, was a radical who would speak to the people. The influence on such liberal men was profound, and the village began to slide toward secularism.

In the summer of 1911, the tavern caught fire. So rapidly did the fire spread that Aba Brown and his family scarcely escaped alive. Arson was suspected, and the police began their investigation, but they could implicate no one. Aba Brown tried to revive his business in the village, but that proved unsuccessful because of its new location. The tavern was so located that only the villagers frequented the tavern, while persons traveling through Khom'yakivka, unlike before, knew nothing about it. Aba Brown tried another location and bought another building from a fellow Jew, who had emigrated to the United States and where he stayed till the beginning of the war, but even then business did not go as well as it did at the crossroads.

With the fire having destroyed the tavern at the crossroads, the village experienced a cultural revival of sorts. There were concerts by the brass band of Sich from Tysmenytsya and plays by a drama group that performed "Dobosh's Death," "Verchovintsy," and "Hritz, Please Don't Go!" They were held in the central square of the village, near the library. Father Durdello was responsible for this surge in cultural activities. Moreover, he would speak at the library about history, people's rights, the Austrian legal system, and Polish oppression of Ukrainians, and speak out against drunkenness.

The constant fighting with the Sichers took a toll on Father Durdello's health, and he was admitted to the hospital, where he, not one for self-pity, became absorbed in thought. Once back from the hospital, he asked those at mass to gather in the library. He wanted to share his ideas with the villagers. The library was overfull. Many villagers stood outside to hear Father Durdello speak. After that event, fewer and fewer villagers came to listen to him speak or preach, and the priest was badly hurt. He was always fully impassioned, caring, and sincere, but his fiery speeches and sermons increasingly failed to move the villagers, the men especially, in the manner he had wished.

In the next several chapters, I describe Khom'yakivka through its principal religious holidays, for the largest events of the year were the religious holidays. [7]

# Chapter 4

## The Christmas Holiday Season

BEFORE THE CHRISTMAS HOLIDAY of 1911, Father Durdello invited girls and young women to decorate the church for the holiday. He, of course, oversaw the decorations. He had a much greater influence over women than over men, many of whom were becoming radicals. The church that year was so beautifully decorated that people from other villages came to see it. There were numerous lovely flowers made from colored paper, and each icon had beneath it a small bucket of roses. The central icon had the inscription, "Jesus Is Born" (Христос Раждається). The effect was enhanced with the flickering light of numerous candles.

The villagers were extremely proud that they had the most beautifully decorated church that year. Not everyone knew that the decorations were chiefly the work of Father Durdello. In sermons that year, he thanked the women for their efforts in decorating. So excited was he that he even thanked those women who did not help in the decorating. All were assured that God would return their kindnesses.

A week before Christmas (Різдво), the air was redolent with the "spirit" of the occasion. There was plenty of work for the villagers. Farmers fed their cattle, cleaned the stables, and in general tried to make everything outside of the houses ordered and agreeable. Their wives and daughters decorated their houses. They bleached the walls with a bit of blue lime, smeared the floor with yellow loam, and drew unusual but ornate designs, some ten to fifteen centimeters wide, on the walls by the stove. Icons were adorned with artificial roses and fragrant cornflowers. Even little children were given certain tasks. There was no time for anyone to relax at Christmas.

On Christmas Eve, everyone, even children, fasted in preparation for the Holy Supper. As night approached, each father, as head of his family, brought in sight a Christmas doll and said to his family, "I wish all of you a very merry Christmas. Let's live at least another—no—one hundred more years! Jesus is born!"

"Praise him (Славіте Його)!" all immediately replied.

Each father would then give the doll to his wife or daughters, who would decorate it with green sprigs of pine and artificial flowers, especially artificial roses. The doll was made of wheat, oats, and a rye sheaf that were chosen during harvest for the occasion. The mother would lay the table and make twelve different courses to honor the twelve apostles. On each corner of the table, there was garlic, which was there to protect each member of the family from disease.

With everyone dressed in clean or new clothes, family members waited anxiously for the first star to appear in the darkening sky.[8] Children, particularly hungry from fasting, were especially vigilant. Punctuality was very important for the Holy Supper because it was believed that being late for that dinner would make one late for all important appointments thereafter.

When the children espied the first star, the family would ready themselves for dinner. The father or head of the family would stand next to the Christmas sheaf to gather the family. He would offer up a prayer of thanksgiving to God for having allowed the family to be together for another Christmas and bless all the food. Then, with a look of dignity and honor, he would say, "Well, dear family, let's sit and have our Holy Supper."

Some superstitious families would practice fortune telling. The family head would throw wheat porridge to the ceiling. Inspection of the number of grains stuck to the ceiling was believed to be a sign of prosperity—e.g., how many sheaves they would have in their field for harvest or how many swarms of bees they would have in their apiaries.[9] We never practiced fortune telling, as my parents believed that real Christians did not do such things. They merely loved God and recognized his kindness.

In the center of the table, there sat two loaves of bread, wheat or rye, each with a lit candle stuck in it. The loaves were called Basil Meals because they were left on the table till New Year's Eve—the day of Saint Basil.

The twelve courses, meatless, began with wheat porridge. Then borsch with herring-stuffed donuts was served, followed by three varieties of *varenyky*, millet porridge, *holubtsi*, cooked fish, potatoes, and sugar beet. For dessert, there were sweet *pampushky* (sweet donuts) and cake. The food was Lenten, except at Pylypivka there was also served pumpkin, hemp, or flax oil.

When the Holy Dinner was over, the family would sing carols—especially "Immortal God" (Бог Предвчний). After caroling, children rushed to their grandparents to receive affection and holiday presents.

Khom'yakivka was alive at this very cold time of the year. People walked around the village and warmly greeted each other as they went to visit relatives or friends. Young people sang carols at midnight to make money, which was often used for painting the church in the springtime, and then attended Christmas mass, which for many of the carolers had already begun.[10]

In the afternoon of Christmas Day, there would be more carolers in the snow-filled streets of the village. The following day, the caroling would subside.

# The Oath

Among the young men, one would volunteer or be chosen to wear a Jewish mask and to dress like a Jew. They would walk around the village streets with the Jew and make derogatory jokes. That generally continued on the second day after Christmas, the Day of St. Stephen. On that day, with the permission of Father Durdello, they might also listen to music till late into the night.

# Chapter 5

## The Yordan Holiday Season

THE PERIOD BETWEEN CHRISTMAS and Yordan (Йордано; Epiphany), which celebrates the baptism of Jesus, was a quiet time—a holy time. That is why adolescents waited impatiently for the so-called Maslenitsya holiday, which was loud and fun—a time in which youths often were married.

The Yordan holiday was similar to Christmas in that it was a time of great celebrating. The eve of Yordan was called the Second Holy Evening—the first being Christmas Eve. There too was a Holy Dinner, comprising twelve courses.

After dinner, girls would rush outside and clang dirty spoons till a dog began to bark. It was believed that the direction of the barking would signify the direction of a girl's future husband. Not everyone believed in such prophesy, but it was an enjoyable shared event for the girls. Meanwhile, women went to church to get holy water to be used for blessing their farms. The rest of the holy water was used to make dough for breads and for small dough crosses to decorate the steps of their houses. The dish upon which the holy water sat was decorated with colorful handmade ornaments.

At around four a.m., everyone went to church for mass.

After mass, each person, holding a cross, went to the chapel, built by my father, in which there was a spectacular, strong spring that became a stream outside the chapel that ran into a nearby river. The spring never froze, even when the temperature was below freezing.

If it was still sufficiently cold, the men would make a cross of ice to be placed by the well in the chapel.

Near the main cross by the chapel, which looked like a small church, Father Durdello would bless water in beautiful bowls held by the women. As the chapel could not hold more than thirty people, many sometimes would have to stand outside the chapel. If it was not too cold, Father Durdello would bless the water for everyone outside the chapel so that everyone could participate in the solemn and beautiful ceremony. After the blessing, everyone, carrying bowls of water and candles, would return home.

Once home, the villagers would place their candle in their house and bless everyone and everything in and out of the house with the holy water. It was believed that the ritual of blessing everyone and everything in and around the house was a way of averting disasters, such as unexpected deaths or fire.

As with Christmas, there would be decorations in and around the house. The outside gate would be decorated with small crosses, made from spikes and red snowball twigs. Some crosses were so beautiful that it was a crime to remove them. Instead, they were left on or above the gate until the Holiday of Meeting (Vernal Equinox), when villagers would finally remove them and store them for the following year.

# Chapter 6

## The Maslenitsya Holiday

THE MASLENITSYA HOLIDAY SEASON[11] (М'ЯСИЦІ) was an especially delightful time for the young. It was a time when they could act as they wished to act, and many were married during this holiday.

Weddings were unique ceremonies with great traditions. On the day of St. John the Christian, young couples, accompanied by relatives or older friends, approached Father Durdello to ask him to preside over their wedding. The ceremony and celebration lasted about three weeks.

The day prior to the wedding, matchmakers, typically older women, would untwist the braids of the bride and her flower girls and braid the hairs together in a symbolic bond. The girls wore a hood, festooned with golden periwinkles—displaying large, violet-blue, pinwheel-shaped flowers against glossy dark-green leaves with gold outlines—in the form of a crown, and the hairs were covered by a shawl. There were a large number of strings spread over the bride's head on the back of the crown. Thus, the bride, dressed in this manner, and her flower girls would walk through the village, visit almost every house, and ask the villagers to her wedding. "Father and mother asked me to invite you for wheat cake tonight and to the wedding tomorrow," the bride would say. She would then bow down and kiss the hands of older people, while she would kiss friends roughly of her age on the cheek. Invited villagers would thank the bride, and the woman of the house would give the bride a gift such as a necklace, which the flower girls would carry. When the gifts became too numerous, the group would drop them off at the house of a relative and then continue the invitations.

## The Oath

The groom would behave similarly. Dressed in a tall woolen hat with golden periwinkles on its right side, he would walk around the village with his best man and groomsmen, with strings draped on the left side of their hats, and invite villagers to the wedding with the same words as the bride used. It was not the custom for them to receive presents, however. They, too, bowed to the elderly and kissed their hands, while they would kiss the cheeks of those of the same age.

After having been in more than two hundred houses, the couple and their party would be so tired that the wedding did not much seem to matter. Still, they were happy, for having chosen to be married, they were acting pursuant to their wishes. Marriages in Khom'yakivka tended not to be arranged.

That night, prior to the wedding, the villagers would come to taste the wheat cake—an enormous and beautifully decorated sweet wheat bread. The matchmakers made wreaths of periwinkle, and everyone would sing all sorts of songs. One song was nostalgic. It was about a beautiful wreath, made by a small girl, which was placed in a trunk for preservation. Years later, the girl is about to be wed, and she takes the wreath from the trunk and remembers when she made it. Another song was merry. It was about making wreaths while drinking vodka with honey and still having a dry throat. The songs were often interrupted by especially cheery songs or the dancing of young men and women.

When the wreaths were made, the bride's mother offered supper to the villagers. The young ones would dance, drink, and celebrate till late at night.

The next day, the matchmakers would come to the house of the bride, sit her in a chair in the middle of a room, and comb her hair. While combing, they would sing very sad songs. Those songs were unique in that they were spontaneously created. Yet one song was popular. It was about young birds flying away from home. The bride would typically cry, especially if she was an orphan or if her mother could not attend the wedding, say because of illness.

When the bride's hair was nicely combed, the matchmakers would place a wreath of golden periwinkles and some garlic on her head. The garlic, of course, was symbolic—to preserve the health of the bride. The bride wore a long, white linen. With the wreath in place, the matchmakers would drape a red shawl with flowers at its top on the bride's shoulders. She was now readied to be married.

With the bride readied, the groom would come. He would arrive on a horse, be carried on a horse-drawn cart, or walk with his best man and groomsmen. The party would sing love songs with instrumental accompaniment. One such song was about a young man who, at first sight, fell irretrievably in love with girl, but the girl did not feel the same way about him.

The bride would wait by the gate of her house with her bridesmaids and matchmakers. When the groom and his party arrived, he would kiss each of the bridesmaids and then enter the house to greet the parents of the bride. Her parents would be seated near the dining table. They would have a loaf of bread in their laps. The young couple would kneel before her parents, kiss the bread, and ask for the blessing of her parents.

"God bless you," said her parents, who would then kiss the young ones on the forehead and hold one loaf atop the head of each of the couple and sing a song about a girl who knelt before her family and expressed her gratitude for everything that they had done for her. The ritual would be repeated thrice. If the bride happened to be an orphan, the song would sadly be about a girl who kneels before others without a reason, as she is without a father and mother.

After that ritual was performed, the party and villagers would make their way to the church while musicians played the wedding march. The villagers would sing such lyrics: "We are walking to the wedding with the bride and groom. Both are beautiful as flowers, sweet as strawberries, and very happy." Or, "Oh, girl, how can I not love you, when you are so beautiful?"

**Picture 6.1. Some Ukrainian Staples**

The couple, the groom on the right and the bride on the left, would hold hands as they walked and carried a red shawl. Musicians preceded them and played various melodies. The couple also carried with them tasty wheat cakes. The groom's cake was tied to his right side by the red shawl, while the bride carried her cake in her left hand. As they greeted villagers while they walked to the church, they broke off pieces of the cakes to share with the villages. The best man also had a wheat cake, tied, however, to his left side by the maid of honor.

A wedding was a special event in the village. When the wedding ceremony at the church was over, the villagers would go to the bride's house for dinner. As the newly wedded couple walked to her parent's house, the groom would now be to the left of the bride, and they no longer carried the red shawl. The matchmakers would sing, "Thank you, dear reverend, for all that you've done," "Oh, my dear mother, I am no longer yours. I belong to the man I just married," or "Oh, the bride, thinking that she would never love her husband, was crying on the way to church, but now that she walks with him, she realizes that her sweet husband is the most significant person in her life." The groomsmen would sing, "This girl once seemed to have such beautiful eyes, but we no longer see their beauty now that she belongs to him."

When the couple arrived at her parents' house, they met her father, who waited by the gate with glasses of honey and bread in his hands, and her mother, also with glasses of honey. The first glass was given to the newlyweds, who shared the honey. The other glasses were offered to other guests—relatives or close friends. Those invited to dinner greeted each other and proceeded into the house, where they sat down at the table. The newlyweds, surrounded by the groomsmen and bridesmaids, sat at a privileged place in the middle of the table, where they and their relatives were served a sumptuous dinner.

During the dinner, there were many speeches given on behalf of the young husband and wife and many songs.

After dinner, there was music and dancing till very late.

When the wedding was over, the groom and groomsmen would take the bride to the groom's house over the bridge on a highly ornate horse-drawn cart. Once there, the matchmakers were already singing. "Good evening, everyone. We came to look for our little goose, which accidentally flew into your yard. Please give us back our goose or take in us as your guests."

The groom's mother, standing outside of the house, offered the guests *horilka*[12] (горілка) with honey. It was the custom of our village, for the groom's mother had to demonstrate ritualistically her authority over the bride. She would take the bride's coat, turn it inside-out, and wear it. Wearing the coat, she would kiss and hug the bride. The bride would kiss the hand and cheek of her mother-in-law. When the ritual was over, she and all the guests were invited into the house.

Once inside, there would be further merrymaking, lasting usually till early in the morning. After some drinks, the matchmakers began to sing, "Our matchmakers bought horses—small, elegant, beautiful horses," or, more humorously, "Our matchmakers are all sick and filthy, and only one is healthy, but even she has a huge pimple on her back."

The merrymaking—the dancing, eating, joking, and fun—lasted till morning.

The partying was, of course, loud, but it was always, or at least mostly,[13] respectful fun, and there was always a concerted effort not to offend the elderly. If a young man got drunk and tried to start a fight, then someone his senior would merely ask him to leave, and he would. Thus, the ceremonies in the village, though loud and joyful, were seldom raucous, and so the villagers always looked forward to them.

# Chapter 7

## The Easter Holiday Season

LIKE OTHER LARGE HOLIDAYS, there was much to do in and around each house in preparation for Easter (Великеня). Villagers cleaned their yards—sometimes two yards a day, as the work was shared—and things that were no longer needed were carted onto the fields and deposited. Stables had to be cleaned. Everything needed tidying. Each farmer offered food and drink to his workers. So important was preparation for Easter that villagers would poke fun of any farmer who failed in the spring-cleaning tasks on Easter Sunday.

The work inside and outside the house was conducted by women and young girls. Many would gather in one of the houses and work together. They would paint the house and embroider shirts for themselves and their husbands while they sang holy songs. Each house, inside and outside, was meticulously painted.

Holy Thursday began the preparation. It was always a noisy but sad day. Each worked in and around his own home. Women baked decorative and tasty Easter breads, cakes, and other foodstuffs. Men tidied up their stables, took care of the cattle, and did the shopping in town. The girls would all gather at one house to make decorative eggs for Easter.[14]

In the evening, everyone dressed finely and went to church to listen to the service in honor of the Last Supper, where there were readings from twelve Gospels.[15] After each Gospel was finished, there would be a ringing of bells, and the bells would remain silent thereafter till Easter Sunday.

Good Friday was always a sad, quiet day. No one worked the fields, and solemnity reigned. No one laughed, whistled, or sang. Farmers attended to unfinished chores, but did so taciturnly. Any communications were conducted gravely and quietly.

At dinnertime, there would be the stentorian beating of wooden sticks on a large, wide wooden board. That was the sign that Christ had just died on the cross. Villagers would immediately stop whatever they were doing and make the sign of the cross.[16] Older women would kneel or bend down and pray. It was to me a beautiful tradition of real Christians, but that tradition would sadly begin to die over time.

When the wooden sticks were beaten on the board for the third time, the villagers left for church. It was time to take out the statue of St. Mary and kneel before the tomb of Jesus Christ.

On Holy Saturday, there was again the effort to take care of unfinished chores. The young men would make a large fire in the yard to be watched carefully all night long. Boys decorated their clothes, and it was the custom to place a peacock's feather on their hat. The women prepared meats and cakes and rolls.

Girls, especially, attended to their appearance. They dressed, minus the long white linen, as if they were to be wed. Each girl's head was covered with flowers and strings. It was with such decorating as it was with embroideries. The number and beauty of adornments were seen as signifiers of wealth. Her clothes for Easter had to be new. Each would wear yellow or red boots, a colorful skirt with an embroidered belt, and a beautifully embroidered white shirt that had dangling strings of varied colors. Poorer girls would wear the clothes worn the year prior by a friend. It was important for richer girls to make certain that their poorer friends dressed as smartly as possible for Easter, so great was their love of and concern for their friends.[17]

**Picture 7.1. Church of St. Michael the Archangel (Today)**

Easter, or the Day of the Bright Revival of Christ, was a big day for all the villagers. They woke early, when the stars were still in the sky, and dressed smartly for the morning service—the blessing of the baskets. Most took with them baskets of food, which were placed around the church, for Father Durdello to bless.

So unique and solemn was the event that it is hard to capture with words. Imagine a sea of flickering candles in the otherwise dark church, quiet singing of the chorus—"Jesus Christ rose from the grave"—the harmonious pealing of bells, and Father Durdello, dressed in a light holiday robe, walking by the villagers and their baskets and sprinkling holy water on the people and their food.

With the service finished, the villagers were greeted by the rising sun and a mild spring morning. They were exhausted from all the preparatory work for the holiday but still happy that Easter had finally come. They greeted each other with kindness and love, kissed each other, and each wished every other a happy Easter. They would then return home to have their breakfast of blessed foods.

After Easter breakfast in 1912 and a small respite, everyone again headed to the church for a daylong celebration. The young adults had already gathered on one side of it. Girls sang songs such as "Zelman," "The Gardener," "Zshuchka," and "Nastochka," which are now widely known all over Ukraine. Young men gathered themselves in a group of sixteen, as if they were soldiers, and from among them, they chose a leader or commander who was a real soldier. The commander would show them soldiery things, such as marching as a unit. The young men then began to march toward the girls, who were then singing "Nastochka," about a lovely girl with beautiful, silky clothes. The girls, afraid of the marching young men, would scream and scatter.

Many other things were happening around the church. On another side of the church, there were gymnasts, who attracted many spectators, amazed at their athletic feats. Other youths tried to circle thrice the church while balancing something breakable. Some lost their balance and merely fell into the large group of other youths.

Older people sat outside the church and exchanged humorous stories or told jokes. Mykyta Semanyuk, also known as Protsak, and Dmytro Sapa, were notorious joke-tellers. Each, it was said, could make even the dead laugh. If a beggar walked by the church, the women, gathered the whole day by the church, would give him a week's worth of food. So joyous was the occasion and so great was Ukrainian generosity and hospitality at the time.

My mother, I should add, was a prime example of Ukrainian generosity. On Holy Saturday, the day before Easter, she would send my younger brother or our maid or me out to give butter and cheese to our poor neighbors so they would have something to eat on Easter. "Go and take this food, and be grateful to God that you have enough to give to those less fortunate." When I was six or seven, I resented having to give away our food to others. I could not then grasp why they did not have the food we had. Yet I always went because my mother bid me to go.

The day after Easter, called Bright Monday, the fun at the churchyard continued. With Easter having passed, the girls would not dress so finely. Why? Tradition has it that the young men got into the habit of dowsing the girls with water. Since the church was atop a hill, the young men would have to carry their water up the hill. It was hilarious. Sometimes a brave young woman would haul up her own water and dowse the guys.

These courting rituals could sometimes get out of hand. On at least one occasion, the boys struck the girls with small sticks. Striking the girls, my father objected, was not to be tolerated, and the tradition almost completely disappeared after World War I. Any boy attempting to strike a girl would be reprimanded by an adult and asked to apologize. If he refused, he was made to leave.

The merriment, though the crowd was less and comprised mainly of the tireless young ones, generally continued on the churchyard on Bright Tuesday. Many young men would dress in working clothes in readiness for returning to the fields the next day. One game was weird, but the same very funny one played at Christmas. Some brave boy would be dressed as a Jew and would parade around the village and tell jokes. The "Jew" at some point would have to catch one of the girls, and hug and kiss her. The girl, repulsed by a kiss from a Jew, would spit and slap the "Jew." If the Jew played the part well, he would be chosen to be the Jew during the Christmas holiday season, unless he would wed before then. The Jew had to be a bachelor.

From Easter till the end of Bright Tuesday, the bells would peal, unless service was being held. In the evening on Bright Tuesday, the pealing would cease. The Easter holiday season was formally over. The young ones, dejected, headed home in preparation both for the hard work in the spring fields, which would begin on the morning of Wednesday.

# Chapter 8

## The Green Holiday Season and Lesser Holidays

THE SO-CALLED GREEN HOLIDAY (ЗЕЛЕНІ СВЯТА), the Feast of Triytsya (Trinity Sunday or Pyatydesyatnytsya; Pentecost[18]) was the fourth-largest holiday. Why was it called green? On Green Saturday, the day before Green Sunday, boys and girls decorated their houses with green leaves, especially leaves of maple and lime. Their gates and front doors were especially decorated. The roofs were festooned with small branches of lime and aspen. In the houses, the floor was covered with branches and leaves of aspen, so that a pleasant aroma filled the house.

In celebration of the Green Sunday, a huge procession of well-dressed people would visit every field in the village. First, there would be children who would ring small bells; and they would be followed, in order, by a prominent villager carrying a large cross, about ten villagers holding icons, Father Durdello, a chorus of young singers, and then most of the rest of the village. Numerous signs with crosses littered the village, as did other religious icons. By each cross or icon or even chapel, the priest would conduct a brief service and leave behind the cross or icon on a nearby field. All in the procession sang along with the chorus. At times,

girls would run into a field to pick the best spikes of wheat or rye to make wreaths to place on crosses. Doing so, they would trample on the spikes. The owners were proud, not angry, that their spikes were chosen.

# The Oath

There was a longer service by the large cross, situated at the crossroads to the village of Pshenychnyky. The cross—placed there on May 16, 1848—was in honor of the abolition of the exploitation of peasants by landowners. On that day, many people marched to the cross.

By the end of the procession, all the crosses or icons left in the fields were adorned with wreaths. The procession would exhaust its participants, especially the boys carrying crosses. Sometimes the weather was bad, as it was on this day in 1912, and so the boys rested in the gardens or by the stables.

They were rewarded the next day with beautiful weather.

Another important holiday was the birthday of St. John the Baptizer. In preparation for this holiday, houses and stables were adorned with black-and-white burdock and orange-yellow flowers called the Blood of St. John, which grew plentifully everywhere in the meadows and grasslands. The latter were of especial significance because, though orange-yellow, they would turn blood red if rubbed.

Villagers were at the time still ensorcelled by a superstition. On the night of St. John's birthday, they believed that witches roamed and stole milk from the cows and all the fern that bloomed in the woods. According to the superstition, if a brave young boy would dare to enter the woods and there see blossoming fern, he would be forever happy. On one occasion, a party of boys entered the woods to see the fern blossoming. Some fell asleep in the woods. Others rushed home to the village, where they reported the most fantastic and frightening stories. Every rustle of leaves, every flight of a bird, was a storm of marching devils. After that event, no one dared to enter the woods to look for blossoming fern on the night of St. John's birthday.

**Picture 8.1. Statue of Mary, Mother of God, Today**

On the day of St. Peter and St. Paul, the villagers went to Pohonya, a small village in the middle of the woods, for confession. Pohonya was situated in the valley between two big hills and the villages of Klubivtsy, Pshenychnyky, and Nadorozhna. In the village, there was a small, ancient chapel that contained a majestic icon of St. Mary and a small monastery devoted to St. Basil.

At the chapel, one can still hear the legend of Boonya, a Tartar chief. A detachment of Tartars was sent after the peaceful inhabitants of the valley. The people hid in the monastery. Boonya followed them. As soon as he got to the monastery, he was blinded by the beauty of the valley and the chapel. The Tartar soldiers, seeing their chief now blind, panicked and ran. On the spot at which Boonya stood when blinded, a large spring of water burst out. That spring turned into a small stream, which ran mys-

## The Oath

teriously between the hills. The stream eventually found the Crow River, a branch of the Bystrytsa River. Those living in the area made full use of this miraculous water.

On the Holiday of the Blessed Virgin[19] (Annunciation[20]), everyone, even those from neighboring villages, went to the cathedral in Stanislavivka, where confessions took place. I have to say that none of the people from our village missed holy holidays. On such days, they would eschew hard work in the house and in the fields.

On the holiday of St. Mary, everyone would go to the small town of Odai, situated by the road Tysmenytsya-Kolomyya (H-10) and between the towns of Markivtsi and Slobidka. There was a very large church of St. Mary, which had a beautiful icon of the mother of Jesus. The women who gathered from various villages made a huge garden of flowers.

So many people from so many villages went there, and there were sometimes even priests from Transcarpathia. As many as ten thousand people would gather for confession, and that number the church could not hold. The number of confessants was indeed proof of the devotion of Ukrainian villagers to Christ. World War I, unfortunately, destroyed the devotion, integrity, and deep faith of our people. I visited this and other places after the war. Fewer appeared to confess, and those who did were not so devoted.

# Chapter 9

## Harvesting the Crops

**Picture 9.1. Typical Ukrainian Wheatfield**
(Courtesy of Pixabay)

HARVEST WAS THE FAVORITE time of the year for farmers. Only a true farmer can imagine and understand the happiness, satisfaction, and pride felt when he, having driven his horses and oxen back to his house, brought with them his yield.

The harvest in our town was special. On a quiet night in late July or early August, the girls and boys would leave their houses and sing melodic songs under the light of the moon. The boys and young men would enter

# The Oath

our fields, even the borrowed fields of the priest, and reap the rye, wheat, and oats by the light of the moon. Then they would tie together the stalks and put them into square-formed heaps so they could dry well. So efficiently and rapidly did the young men work that in just a couple of hours there were numerous squares of wheat. My father often went out to watch their machine-like efficiency for hours.

When the moon was about to settle and dawn was approaching, the young men—with their "Princess of Crops," a young woman who was designated princess by them—returned to the village.[21]

In our house, mother and sister greeted them with bread and honey or cheese and sour cream. The princess would remove her wreath and place it on my mother's head, and that signified the end of the harvest. There was *horilka*, Ukrainian vodka, for the young men, who gladly drank it, though each was offered only one glass.[22] Then there was dinner.

This celebration often lasted past midnight, thus father permitted music played outside so that the young ones could enjoy themselves till the morning. Our neighbors never complained. Instead, they often participated in the celebration.

Throughout the night and early morning, my mother offered a steady stream of cheese, sour cream, and honey; and for the young men, there were two liters of *horilka*. Father gave the *horilka* to a responsible middle-aged man who was to make sure that no one would get drunk, so that there would be no rowdiness, only fun.

In the morning, the young partiers were exhausted. The music ceased. There was only conversation and snacking.

The process of taking the heaps of crops to each house was similar to harvesting. Each farmer had his own helpers from the village, some of whom would even bring their own carts to help take the harvest to the house.

# Michael Chemny

I remember vividly those remarkable days. A huge cart would drive on the dusty road, and the owner, holding his whip above his horse, would sit high on the heaps. He would whip infrequently—only enough to show the horses that he was their king, sitting high atop their food. The horses, knowing that they were pulling food that they would later get, obeyed and pulled the massive load. It was hard for anyone, seeing the pride and satisfaction of the farmer and resolve of his horses, not to feel the pride and satisfaction of the farmer—so obvious and infectious were his feelings.

I recall one year as a very young boy, when I rode on such a cart, overfilled with square heaps, and it is now impossible to describe how I felt. When still in the field, the ride was bumpy and slow—the horses struggled. When we managed to get to the road, the horse would start to trot. Once home, someone took me down from the cart. Men loosened the rope holding the sheaves and placed the sheaves in such a pattern to keep most of the crop dry during even the wet autumn season.

# Chapter 10

## Mowing after Harvest

**Picture 10.1. Reaping the Harvest**

(Courtesy of Pixabay)

# The Oath

AFTER THE HARVEST, VILLAGERS gathered for mowing the wheat. It was again a very pleasant time. Early in the morning, they went into the fields. Each took off his hat, made the sign of the cross, and said a short prayer in which he asked the Lord to help him during the arduous day. After the prayer, each mower would sharpen his scythe, fill his *kooshka* (a woken cup affixed to his belt), and begin mowing. [23]

Soon one would hear the melody of what seemed like a million scythes—the music of metal. That music was interrupted from time to time by the voice of one of the music makers, who would ask another mower where he was intending to mow.

The village had two large fields—Voonyava (110 square kilometers) and Krichman (81 square kilometers).[24] Voonyava was the most beautiful, as it was surrounded by woods, and so it was mowed before Krichman. The name Voonyava came from a river on that territory. No one knows how Krichman got its name. It was some nine kilometers long and wide. Not surrounded by woods, its hay dried earlier.[25]

Mowing old grass is difficult, exhaustive. Mowers were typically drenched in sweat, and they had only a half of an hour for lunch, which was brought to them by a wife or child.

When the sun set, mowers would put away their tools and thank God again in a prayer for helping them through the day. The mowers would then gather and begin to sing as they headed home. Perhaps because they were the result of a hard day's work, the songs of the tired mowers were typically charming and provocative. Their voices seemed to me so majestic, clear, and sonorous. The songs and the manner in which they were sung echoed joy and melancholy, happiness and sadness, satisfaction and misfortune. It was as if those feelings were alive and trying to leave the souls of those tired workers. The singing was also a sign to everyone to finish their work in or around the fields. Children led the cattle home.

Girls and women hurried home to make dinner. How beautiful was village life at that time!

In two to three weeks, haystacks appeared on the hayfields. It was not easy to make haystacks. Using their pitchforks, mowers had to arrange the hay in such a way that mounds would not take in too much water when it rained. Hence, they shaped the hay into large mounded hills.

The holiday of Holy Preobrazhenya (Feast of the Holy Transfiguration of Christ[26]) followed the mowing. Father Durdello would bless flowers and vegetables, in baskets held by beautifully dressed women, in a special ceremony in the church.

# PART II
# WORLD WAR I

# Chapter 11[27]

## Tragedy Strikes

THE YEAR WAS 1914, just before Easter break. I was just fifteen and called from school and asked to go home. My mother was dreadfully ill, and Dr. Yanovich recommended immediate surgery. He sent a letter to the surgeon, Dr. Soloviy of Lviv, and bid him to take great care of my mother.

My father planned to go to Lviv with my mother. The rest of us—I especially, since I was the eldest—were supposed to stay on the farm and take care of it. That was especially difficult since our farm overseer of seventeen years—he would look after the cattle and generally manage the farm—had recently married and removed to another little town. That bothered my father little, as the overseer was generally lazy around the house, and he often would, to the dismay of my father, stupidly and dangerously sneak into the straw-filled stable to smoke. With the overseer gone and with my father going to Lviv, management of the large farm fell on me, young and inexperienced and only fifteen years of age. It also fell on me to be a parent to my three younger siblings.

It was a difficult time. We were helped by our aunts and uncles, who did what they could, but the lion's share of the work was mine. Moreover, the villagers who frequented our farm to visit my dad to ask his advice on all sorts of matters did not visit my father in Lviv. Though my father always did what he could to help the local villagers—he shared everything that he had with them and stood up for them when they were in trouble—they abandoned him in his time of need. The hypocrisy was plain.

# The Oath

Dad remained in Lviv for three and a half weeks to take care of Mother. When they returned, we were ecstatic. Mother, still very weak, had to remain in bed all the time.

As it was still winter, our chores as kids, under our mother's supervision, were confined mostly to the house. Yet spring was near, and it was now time to return to the fields. Father, a builder, needed to resume his job. Having spent over three weeks in Lviv and away from his job, his pockets were empty. The hospital bill was large.

One day, after Dad returned home from work, he asked me to come to his room, where he said, "Son, you won't attend high school anymore, because I can't afford it. But I still want you to have higher education, and that's why I've asked Judge Kysylevsky to help you to go to Kaiser Shullye in Vienna. You can get a good education there. You'll only have to pay for your place. What do you say?"

Flabbergasted, I did not know what to say. Vienna—*it's not just a city, it's a capital. Strange people, a strange language.* Those thoughts flashed through my mind like lightning.

After some thought, I replied, "Okay, Daddy. I'll go to Vienna, though—though what I really want is to stay *here* with you. You need help here."

"Don't worry," said my father. "We'll be all right here. And you—you better go to work on your German and Latin. You'll be needing both."

What could I do? In obeisance to his wishes, I kissed my father's hand and continued my work around the house.

And so I would be off to Vienna in the fall.[28]

During the day, I worked on the farm. At evening, I read books. When I had problems with my studies, I usually went to see my old teacher,

Maria Blyokivna. Yet she was often busy or sick, and so then I would go to Father Durdello, who was generally glad to help me. He even gave me some badly needed books.

Spring was beautiful that year. It was mild and with the right amount of light rain. The villagers, who seemed to be everywhere, were anticipating a good harvest. They worked our field too, though, as usual, they worked it quickly and hastily. Still, my father treated them well. He even hired musicians for the young so that they could dance and enjoy themselves on Sundays. When they would leave, they wished my father luck and promised to come and help with the harvest. My father was very grateful.

My father located an additional fifteen acres of arable land that he bought and sowed with oats and wheat, through the help of a German book, *The Country Man*.

"You'll have a great harvest this year, Mr. Chemny," said a fellow farmer, old Fedir Zayachuk. "When we walked past your field today, I saw your oats. Well, I must say, it seems like *our* soil! And the wheat? Why, the wheat has gotten as green and as tall as wheat can get!"

Then ever-doleful Nicolas Sheshorak, another land-owning farmer, answered with evident reproach, "It's too soon to say."

"Oh, Nicolas, Nicolas! I am an *old* farmer, and just a look—just *one* look—is enough for me to know if something will grow or not," replied Fedir.

"Enough gibberish, good farmers!" said my father. "It would be great if each of us would *harvest* the gift God has given us, not squabble about it." His words, as they often did, were like a knife's wound to my heart. He seemed for some reason to predict some misfortune just around the corner.

# The Oath

The conversation continued, and now a third farmer, Yurko Mykytyn, had to have his say to my father. "I didn't think that *anything* would grow on the spot you bought."

The farmers often conversed in such a manner with my father, especially in our house. There they discussed business, freely shared thoughts, and even talked about their dreams and jealousies. When Father got tired of their conversation, he led them to his library where he would read something to them—as they were illiterate—for their enjoyment or edification. They were especially excited when he showed them his new, thick book, *The Country Man*. From that, Father promised to read to them something that might prove useful for their farms. Each of them still worked the fields, as did their parents, with a wooden plow, instead of the iron ones in use by the more science-minded farmers.

# Chapter 12

## World War I Begins

ON SEPTEMBER 14, WHICH was a Thursday,[29] I came to visit my father, but he was not home. In the middle of the night prior, every soldier left for Tysmenytsya. They said there that the Russians were already in Nyshnev, but they were closer than that, because at around one p.m. they arrived in Tysmenytsya.

I, too, went to Tysmenytsya.[30] The town was mostly abandoned. There were no authorities and no police. All the shops were closed, and Jewish shop owners, scared to death, periodically peeped through the windows of their shops to see if anything was about to happen. Some farmers, ignorant of the situation, were at the market to sell their goods. At around one thirty p.m. on the road Nyzhnev-Stanyslavov, which passed through the market, there appeared a Russian Cossack,[31] dressed in black and wearing a tall hat. He held a rifle and was ready to use it, if needed. His horse, walking slowly though majestically, was also black. The farmers at the market, curious but fearful, turned to their unexpected visitor. Behind the Cossack, there were two more Cossacks, and behind them, five more. All looked the same; all were dressed the same. Seeing the frightened faces, one Cossack said, "Don't be afraid, people. Be calm."

Thus, they passed the market and headed toward Stanyslavov. Behind the five Cossacks, there was a squadron of Cossacks. When they entered the market, they alighted and began to break the windows of the shops. The case-windows of Zister's shop were broken and all the jewels and items of silver and other precious metals were thrown upon the road. One Cossack shouted, "Take them! They're yours!" The people, afraid, took nothing, but merely stepped back. The Cossack continued: "Take them!

# The Oath

Why are you standing there? What are you waiting for?" He then threw a silver chalice into the crowd.[32] Someone picked up the chalice and hid it beneath his coat.

More Cossacks arrived later and—using iron rods, sticks, or their rifles—they broke up the shop of Vainravh. His store was neatly decorated in the American manner, where all the items were neatly places on unique small boards. They did the same to all the shops on the road and threw all the items on the road. I watched this piracy of goods and was disgusted with what I was witnessing. The people of neighboring villages quickly arrived to pick up the goods. How they arrived so quickly, as many were a few miles away, I cannot say. There was such chaos in Tysmenytsya, and it was difficult to understand what the people were doing and why they were doing what they were doing. I became angry. How could Ukrainians be so easily provoked to steal from each other?

Around three p.m., the Russians set on fire the village pub. There was a stream of *horilka* that ran from the pub, and people came with bowls or buckets to scoop up what they could of that stupefying vodka. Those not lucky enough to get horilka near the pub scooped what they could from the stream on the dirty street. The scene was fascinating: it was comical and disgusting at the same time.

**Picture 12.1. Russian Soldiers of World War I**

The stream of *horilka* finally caught fire, a running blue flame, which eventually might prove a threat to the large wooded bridge afar and on the main road. An officer bid some of the Cossacks to stop the fire and save the bridge.

Sickened, I began slowly to walk home. As I walked, a stranger said to me, "Aren't you going to take anything home?"

"No," replied I. "I'm not used to taking what doesn't belong to me."

"Oh, how stupid you are!" stated the man, carrying a large sack of stolen items on his shoulder.

# The Oath

It was late when I got home. There was pillaging in Khom'yakivka, too, though there were no Russians in our village. Father Michael Durdello, a true patriot of Ukraine,[33] was afraid of being imprisoned by the Russians. He gathered his belongings and his relatives on his carts and left for Austria. His father-inlaw, Dean Kalynsky of Chornolozets, did the same.

The farmers, saying that they wished to save what the priest did not take with him from the Russians, took everything from the house and yard—even his cattle, his chickens, and his geese. Some villagers tried to break into his stable, but they did not have the keys to it, so it was safe from theft. The priest also had a small mill on the Strymba Stream, which was east of the village and ran through the southern part of Khom'yakivka. Villagers pillaged that mill.

The thievery was horrific! People stole huge barrels of honey, which they placed atop carts or merely carried on their shoulders. One man, carrying a large barrel of herring and trying to get across the stream, slipped and fell in the stream. Most of the herring went into the stream. Some villagers stole his grain; others, his flour. The priest's valuable library was plundered. Around his house, in his yard, and even everywhere in the village, there were torn pages from his books, magazines, and newspapers.

Our village looked like it had suffered a terrible storm. Back in my house, I was given the day's grim details. I walked to the hill, across Father Durdello's house, and stood next to a large cross. There I examined the priest's farm. So terribly was it looted that it was unrecognizable. I was much ashamed of my neighbors' wickedness.

One illiterate fellow was carrying a stack of the priest's books. "What are you going to do," I asked, "if Father Durdello should return tomorrow, or the next day?"

"So what!" he replied. "Would it be better for the Russians to take them? Anyways, as soon as he returns, we'll—we'll return everything."

"Yes, but what about the things you've ruined? Who is going to give back them?"

He said, "Then—well—*you* go and take the things that are ruined to save them."

Filled with indignation, I returned home. It was an unthinkably dreadful day, and I was exhausted, so I soon went to bed.

At around nine thirty p.m., I was waked by the barking of dogs in the village. *The Russians have come*, I thought. I threw a coat on my shoulders and ran outside.

"Thieves, give it back! It's not yours!" said Dean Kalynsky, who returned to the village when he heard of the plundering. He had been driving slowly on the road that led to Chornolozets. As he drove, he grew angrier and angrier. Would his old chest take the pressure? His heart could give out.

Dean Kalynsky had everyone in the village on his heels. Those who stole the priest's cattle and horses returned them that night so as not to be seen. Thus the chickens and geese, too, were returned. However stealthy they tried to be in returning stolen items, the dean would catch them and give them a considerable piece of his mind. Dean Kalynsky did not go home that night but spent the night at his son-in-law's house. He slept on the filthy floor because the "good people" even stole the bed.

Two days later, Father Durdello returned. Having seen the devastation to his house and especially to his highly prized library, he wept bitterly. It was Sunday night. He did not go to the church to pray. He had not the strength for that.

Greatly saddened, he went to the small chapel, built by my father, and prayed by himself. Many farmers watched him. None had the courage to approach him. I did. I knelt in the corner behind him, and I, too, prayed.

# The Oath

Father Durdello turned and saw me, but he merely turned away and continued to pray. I wanted to leave the chapel, but I did not. The priest then stood and turned to me. I bowed. He came to me and said in a moan, "Mickey, it's horrible!"

"Forgive them, Father, for they know not what they are doing."[34]

"I have already," said he, but the sorrow and sadness were easy to see in his eyes.

The villagers tried to return what they stole, but so many things were not returned. Some were just too embarrassed to return what they stole. Other things were returned at night. Father Durdello had no grain and no flour to get him through the winter. With the return of their priest, there was a feeling of panic in the village.

I was angry and ashamed. Our villagers, after all, were not such good people. They were bad—bad and irresponsible.

Years after the war, I reminded my fellow farmers in my reading room about that time our priest was robbed by his own parishioners—the farmers in the reading room. My words brought them pain and shame, but they countered that they were merely trying to save his property from the Russians. To me, these farmers were worth less than the Russians.

Father Durdello would never recover from the plundering. During the war, he became increasingly poor and apprehensive, as his sons were in the war.

We, too, suffered. We had only one horse, so that is why we had difficulty when we had to travel to Stanislavivka or to Tormach.

The Russians then came and remained for two or three weeks. They left when the front moved to the Carpathian Mountains. Sometime after the Russians left, a group of armed Austrian horsemen road through our village. They somehow got behind the Russians and rode the dusty roads to Tysmenytsya, where they killed a Cossack and fled into the woods.

That event upset, even panicked, the Russians, and they ordered some cavalrymen and foot soldiers to find the Austrians. The search was brutal and terrifying. Like uninvited ants at a picnic, Russian soldiers were everywhere, and like ants, they destroyed everything on their path. They turned over corn heaps, tore up potato fields, destroyed stacks of hay, and ruined our oats. That devastation continued for two weeks, and we suffered mightily as a consequence of the Austrians' actions. We were never told whether the Russians found the Austrian for whom they were looking.

In the winter of 1914–1915, the Russians disappeared, and the Austrians appeared for some two weeks. In that time, Hungarian regiments, in pursuit of the Russians, hanged many innocent Ukrainians—especially those who lived in or around the mountains. They were baselessly suspected of aiding or spying for the Russians. Some, illiterate, could not have been spies. All were innocent.

A post arrived from my father, who had been forced into the Russian army. Fighting in a large battle, he was wounded in the leg, but he recovered and was situated as a border-guard. The post told us nothing else. There was no date, no address. Still, we were happy to be informed that he was alive.

Soon the Russians returned and pushed out the Austrians.

In the spring of 1915, the so-called Gidenbourg's Regiment forced the Russians from our territory and moved them to the east. We could not tell what might happen the next day or the next week, so volatile was the situation. Our lives only got worse. The Austrians stole what they could.

In the summer, there was a general mobilization of Austrian troops. On July 22, it was our turn to be mobilized.

# Chapter 13

## Two Misfortunes, One Day

EVERYONE FROM KHOM'YAKIVKA, on a very hot and very windy day in July 1915, left their houses to send off our Ukrainian recruits, who were then sent to their appointed regiments. Every available man was taken. Those injured in the war were immediately reappointed after having recovered in the hospital. Only the crippled were left behind. There was no house in the village where no man was called to fight.

At around noon on the same day, Michael Krook's house caught fire. With the wind from the northeast, the fire reached the property of Paul Prosnyook and then his neighbor's, and his neighbor's. Within an hour, many houses were ablaze. There was no one to stop the fires. The available men or boys, who were not recruited for the military, were in the fields. Others of my age and I found a fire extinguisher, and we put it to use. But it was not strong enough to put out any fire, and we could not get close enough to the fire to use it. The wells, full of old sticks and rags, too, were ablaze.

Soldiers on carts came later to help us. Putting their horses in a safe place, they helped us to break the fences and gates, which—catching fire—were carrying the fire from place to place. The wind pushed the fire over the stream, and two other farms caught fire. Neighboring villagers from up to four miles away rushed to assist, yet when they arrived the fire had run its course. It ceased before the alder-tree forest in the east.[35] The houses that caught fire were annihilated. Only a few smoldering posts remained.

# The Oath

The spirit of the villagers was destroyed. The great fire decimated the people of Khom'yakivka. It was, in effect, the last straw, as the villagers repined over the fact that hundreds of men, many of whom would meet death, were forced to defend a cause that was not theirs.

Everyone—those who lost a loved one in the war and those who lost everything in the fire—cried desperately. With the souls of the villagers broken, the villagers once again became good Christians and remained pious for some time.[36] Those now without a home took refuge and found understanding in the houses, and hearts, of other farmers. When all the farmers had returned from their fields in the evening, they carried buckets of water to pour over trees still simmering to prevent the fire from rekindling. Everyone paid special attention to Father Durdello's house and yard, which were not far from the territory affected by the fire.[37] We also wetted the houses, stables, and threatened crops of farms near the fire so that they would not catch fire. The mayor of our village had the guard watch over Khom'yakivka that night.[38]

I was then still young, just sixteen, and full of energy, but I was nonetheless exhausted at day's end. I had had nothing to eat all day and too little to drink.[39]

My mother—who had seen early in the day her youngest brother, Michael, off to war—came home late at night. She was so upset and depressed that she could not make supper for the family. We drank to slake our thirst and ate some bread and went to sleep.

After the conflagration and with all the men, from young to old (up to forty-five years of age), stripped from the village, all of us felt for some time as if we had experienced a large funeral. Our faces were joyless, drained. We were depressed. We spoke to each other only when we needed something. The melancholy was contagious.

Women were heard to say, "Who's going to work our fields?" "Where will we sleep tonight, tomorrow?" "What are we, now poor, going to do?"

There were numerous other, similar questions—expressions of misfortune, sadness, and grief.

Though everyone could hear the voice of despair of the poor farmers, no one could soothe their feelings of despair. Father Durdello, through profound and wise words, tried very hard to offer consolation, but that proved unavailing. Our people internalize their misfortunes the wrong way. Perhaps it is impossible to appease the suffering soul of a farmer. Our people are unhappy unless they see that their lives are getting better, easier.[40] Yet unless they see life through the eyes of a sincere but poor person, they will never understand the meaning of "better." All the talking in the world will not matter.

# Chapter 14[41]

## Third Moscow Invasion

THE MUSCOVITES ONCE AGAIN paid us a visit in the spring of 1916. The 296th Berezniki Regiment occupied Khom'yakivka and oppressed and demoralized the people.

In autumn of the same year, the regiment moved eastward. Before removal, they loaded ten carts of goods from the village and took me with them as a guide. I was for months beaten, and the torture lasted until the end of January 1917. They also took my horses, which they exhausted and thus killed. One broke his leg when he fell on the ice, and another became sick. I would eventually get another pair of horses, which I took from some yard in Terebovlskogo County.

As I was constantly with the Russians, I quickly learned Russian so well that I could readily have been thought to be Russian. Yet at the end of January 1917, I was released and given permission to return home. Because my clothes were shabby, I was outfitted in the clothes of a Russian soldier.

It was only with great difficulty that I found my way home. Yet I could not escape the Russians at home, for our house had become the office of the 1st 177 Light Artillery Battalion, commanded by Colonel Dudinsky. When I entered the house, my mother looked at me, clapped her hands, and cried, "Oh, my God!"

I greeted my mother, my sister, and my two brothers, who began to help me remove my snow-covered clothes. I asked the elder of my younger

## The Oath

brothers to unhitch my horses and take them to the stable, where they would have something to eat.

Russian clerks, sitting at our table, looked at me with astonishment. Why would a young Russian soldier not greet them? My mother said to them, "This is my oldest son, Mykhailo."

I walked up to him and said, *"Zdravstvuyte"* (hello). The senior-most clerk, an ensign with a proud face and steady countenance, came up to me, held out his hand, greeted me, and said, "That's good of you, soldier." I learned later that he was a former industrialist of German origin with a high education and good manners and was a native of the city Vyatka. The other two clerks then greeted me, as did their boss.

Mother heated some milk, which I, now seated at the table, gulped down.

Mr. Fominykh—that is what everyone called the ensign—sat down next to me and asked me where I was and how I was doing. He added that, while I was gone, my mother was very worried about me. I was touched by his sincerity, and he was impressed by my command of Russian. We quickly became friends.

There were two typewriters and a telephone.[42] I really wanted to learn how to write on a typewriter, so I asked Fominykh to allow me to use it when his clerks did not use it. He agreed and even gave me a reprint of various reports and records. In some three weeks, I learned how to type.

Once, while I was typing, Colonel Dudinsky and Captain Sukholov, Commander of the First Battery, stopped over. I was working on a translation of a small book, *The Siege of the Mill*, from Russian to Ukrainian. The colonel approached me, took the book, examined it, looked at my writing, and asked, "What are you doing, young man?"

I got up from my chair. "I'm translating this little book into Ukrainian," I said quietly.

"You are a new Gogol?" he said with a smile.

"No, I'm still far from the master."

"Well, well! Can you come over to my apartment tonight?"

"I can," I, a little embarrassed, answered.

"Goodbye."

I was perplexed. The colonel left, and my eyes now fixed on the face of the warrant officer, Mr. Fominykh, to whom I said, "Do you not think, Mr. Fominykh, that I brought you trouble by using your machine for my edification?"

"No, he's a good man. He loves young people," replied Fominykh with a smile.

"Why do you think he invited me to his apartment?"

"For tea, maybe."

"Are you joking?" said I, incredulously.

"No, it is possible," replied Fominykh.

When evening came, I went to the colonel's apartment. I was frightened and perplexed. What would a Russian colonel want from me? The entrance to his room was at the side of a large house, by a garden, so one could visit him without disturbing anybody else in the house.

I knocked on the door. A soldier said, "Come in."

I entered the room and found the colonel was on a chair next to a small table with a large table lamp with a white shade. There was a shadow of

## The Oath

his face on the wall, and the light of the lamp made his beard *(shpitz-beard)* seem whiter than it was. He looked like a professor, not a colonel.

I bowed and greeted him. "Hello, Your Excellency."

"Welcome! Please, be seated," he replied, to my astonishment, in Ukrainian. "I am very pleased that you came." He rang a bell that was on the table, and there came a soldier.

"Tea, please."

"Yes, sir," replied the soldier, who quickly turned and left. A few minutes later, he emerged with a sparkling samovar, two cups, some bagels, and sugar on a plate. He put everything on the table and left.

The colonel poured two cups of tea and asked me to sit down near the table.

I gratefully moved in closer, while he passed me the cup of sugar.

"Please," he began, "I have no mistress to greet and serve guests, so we will merely act like soldiers."

"Thank you! Good manners are good manners."

"Are you a student at a gymnasium?" he said somewhat hesitantly.

"Yes," I replied, "but war got in the way of me finishing my studies."

"Ah, I see. Many of our young people left because of the war."

As we ate bagels and sipped the tepid tea, we discussed many things. The colonel wanted to know where my father was, what he did before the war, what subjects I studied in school, and even what kind of future I envisaged. He seemed interested in everything that concerned me.

He then asked about the village, Khom'yakivka. "Has the village any Moscowphiles?"

"No. Not now. Never."

"So, it's a Ukrainian village to the core?"

"Absolutely," said I.

"And your priest? He, too, is Ukrainian?"

"He's Ukrainian—a great patriot and a good person," I said proudly.

"I met him, but I didn't tell him that I am also Ukrainian, because I didn't know his situation, though, with his position in the house, I can see that he would have to be Ukrainian." He paused. "We Ukrainians need to be careful."

"Excuse me, Your Honor—" I began.

"Colonel, only," he interrupted. "'Your Honor' when there are people around. When we are by ourselves, you can dispense with 'Your Honor.' What else would you like to know?"

"I wanted to ask, if I may—I wanted to ask where you—*why* you, sir, reached out to me with such confidence."

"You are an interesting young man! Translating a book into Ukrainian. And when you're done, would you bring the translation to me? I want to see how it turns out."

"I'll give it to you as a present. It's no great scholarly work. I just wanted to practice on the typewriter."

# The Oath

"Thanks, but I'm not comfortable with the gift. I just want to see your translation. You know, I love young people—especially those who use every opportunity to learn something. You're one of them."

"I appreciate much your kind words. I never expected someone like you praising someone—someone like me."

We continued to talk for some time on different topics, but at some point, I got up from the chair, bowed to the Colonel, and said, "Colonel, can I go? You have a job, and I am wasting your time."

"Why do you think that, Mykhailo Kornylovich? Please, it's my pleasure to speak with you, because it's been a long time since I spoke in my native language, even in my favorite Kyivshchyna. There have been many years of wandering and hardship." He paused. "But tell me, have you in school studied the history of Ukraine?"

"Of course."

"Systematically or selectively—you know, randomly?"

"Systematically—as a subject separate from others."

"That's good. I much like to hear that. When the time comes, you will know who you are and whence you came."

"Well, I—I hope so," I said sheepishly.

"We must be prepared, mustn't we?"

"And so, what news is there about home—the homeland?"

"The news is not good. In fact, it stinks," he said in a hushed voice.

"May the Lord have mercy on us and send us back the freedom we've desired for centuries."

The colonel continued. "Today, it is difficult to say, because—because we don't know what tomorrow will bring us. But we are ready for anything! That I know." He grabbed another bagel. "Please, have another cup of tea." Without waiting for me to reply, he poured me another cup of tea.

We drank tea, talked, and ate delicious bagels for many more minutes. It was getting late. I got up, thanked the colonel for his hospitality, and said, "Colonel, it was very nice to visit in your polite society, your comity, but I have to go home."

The colonel got up and cordially gave me his hand, which I sincerely shook. "And remember, I want to see your translation when finished! Let me know. Okay?"

"Okay," I said, as I bowed to him and left the room.

On the way home, I could think only of the great patriot, whom I met from my beloved Ukraine; and though he wore the golden epaulets of a Russian officer, he did not forget his people and his native language.

When I returned to the house, Fominykh asked me how my visit with the colonel was.

"You were right. We had tea—together."

"Well, you see—and you were afraid. I told you. He's a good man! Really good! He's not just a good conversationalist but, more importantly, he's a good fellow—a likeable fellow."

"Yes, indeed."

The next day, I, with battery charged, returned to my translation and worked with focus. I worked throughout the day without rest. I told my mother that the colonel wanted to see my translation. I said to Fominykh, "I want to finish quickly so that I can return the typewriter to the colonel. Because, well, maybe later he will need it."

In ten days, my translation of *The Siege of the Mill* was finished. I rolled it up in a parcel and left to see the colonel.

In the yard, I met Father Durdello. "Are you here to see me, Michael?" he said with his customary tenderness.

"No, Father. I am going to see Colonel Dudinsky."

"Hmm, what kind of friendship have you begun?" he, curious, stated.

"He is one of us. He doesn't wish the Muscovy to know who he really is."

"I knew something was up, because when he came to my house for the first time, he said 'Very nice' when he saw the portraits of Shevchenko and Khmelnitsky. He is very polite to us—polite and respectful."

"I have had a conversation with him," I said. "And this bundle of paper piqued his interest."

"What do you have *there*?"

"This is a translation, *my* translation, of a small booklet from Moscow into the Ukrainian language," I said with pride.

"How—how did *that* happen?"

"Well, he was on inspection a few days ago, and he caught me at the typewriter and wished to know what I was doing. When I explained what I was doing, he bid me to bring to him the finished translation." I opened the parcel and showed Father Durdello the translation.

"Well, it's very beautiful and looks—looks masterly! Yes, work—work and learn, because now it's necessary—necessary, more than ever. Well, go—I don't want to keep you."

"Thank you, Father." I bowed and soon found myself at the side door of the house that led into the room of the colonel. I knocked at the door and heard a familiar voice.

"Come in!" The attendant soldier let in me.

The colonel was morose. He asked me to be seated.

I thanked him and sat down on one of the chairs in the room.

"Well, do you have good news, Mykhailo Kornylovich?"

"No, it's all the same," I replied.

"What do you have with you? Can the translation be ready so soon?"

"It is," I stated proudly. I unrolled a scroll of paper and gave it to him.

The colonel took my notebook, sat down close to the lamp, and began to read.

He read with passion, and I, excited, stood next to him to watch him read. It took him a few minutes to notice me at his side. With tear-filled eyes, he looked up and said with some irritation, "And what are you doing, Mykhailo Kornylovich? Standing as if at the exam? Sit down, please!"

I was embarrassed. "You know, I was so excited about you wishing to read my translation that I did not notice that I was standing by you. I sometimes get carried away." I sat down, and he read several more pages. "Well,

well," he said from time to time.

After a while, he put down the notebook and told me, "You, Mykhailo Kornylovich, can be a very good writer. But in order to create, you still have a bit to learn. However, in this translation, you have captured not only the content but also the form of the author. You have a beautiful command of the Ukrainian language. You honor your school! I know this story very well, and I can recite it without any script. Hmm, your translation is very good. Where did you learn Russian so well?"

"I lived with Muscovites for more than six months."

"And in such a short time you mastered Russian?"

"Well, I'm grateful, Colonel, that they had me only for six months."

The colonel laughed heartily. "Well, my young man, congratulations!" He extended to me his hand—soft, yet boney and Cossack-like. "Please continue to write. You have a gift."

Appreciative, I shook his hand. "I thank you, sir, for the compliment and advice, but in the midst of our circumstances, I don't know how I can continue to write."

"Well, try. Don't quit. Look forward—look to the future."

"Thank you again, sir, for your kindness and honest words. What our future will be, only God knows."

"That's true, but follow your path. Work for God and for Ukraine. Avoid pride, for pride leads a person to destruction, even in captivity. You know our history—a proud one. Don't you see that our pride was a hundred times punishment to us?"

"That's true."

He gave me the notebook and again shook hard my hand. "When you have free time, please visit me and cheer the soul of an old Cossack who is reluctantly in the uniform of a Muscovite. There are terrible times ahead. Goodbye!"

Colonel Dudinsky clicked his heels. I bowed. We parted.

# Chapter 15

## Revolution in Russia

**Picture 15.1. Russian Revolution of 1917**

TIME FLED. SPRING APPROACHED with another year's worth of work for each farmer. My brother and I were taken to the field, Obernik, to ready the plows and harrows, and to attend to all other farm-related chores for the spring.

In the evening, I sat with the Muscovites in my house, and we talked about many things. They told stories and ridiculed each other. Ensign Fominykh was, as he usually was, quiet—quiet and melancholic. He listened half-heartedly and occasionally added a tart word here and there. [43]

# The Oath

One day in late February 1917, Fominykh received from home a food parcel. It had cookies and many treats. There was also a letter from his wife, which he hid. The goodies, however, he gave to everyone. He kept only one bun.

When his clerks had left the house, he took the letter from his pocket and began very carefully to read it. His face betrayed grim news. Having finished the letter, he threw it into the oven. He then sat on a chair and glumly sunk his head in his hands. It remained there for a long time. When he finally lifted his head, I said, "Has there been some family misfortune?"

"Thank you, Mykhailo. Not yet, but it may soon happen."

"What do you mean?"

"Things are bad. There is a revolution in Russia. The government is threatened. Hungry workers are demonstrating in St. Petersburg."

"Well, that has happened many times and has each time been suppressed, so why worry?"

"Not so easy now. The people are exhausted by war. Working people are under the spell of a socialist reformist. Soon there will be no order. The military front, too, is weak—almost nonexistent. It will be bad."

"Don't worry, Mr. Fominykh. We shall have to suffer through the bad. In my heart, I rejoice. Perhaps Ukraine will no longer be a prisoner of Russia."[44]

In two weeks, on the night of March 17, the phone rang. Fominykh said that he had expected a call at any minute, and he jumped to the phone. The caller reported that in St. Petersburg there was a full-scale revolution. Tsar Nikolay II had abdicated, and Russia had a new government led by the socialist Aleksandr Kerensky. The caller bid the soldiers to remain at their posts because, revolution notwithstanding, there was still a war. The cunning German replied, "Who says?"

"Colonel Dudinsky," said the caller.

Ensign Fominykh relayed the contents of the message to his clerks. One of them, Nikolaev, excitedly cried, "Hurrah!" Another, Korkin, was anything but excited.

The First Battery gathered in the streets of Khom'yakivka to discuss, each soldier in his own way, the unanticipated events. Then came the Third Battery—a crowd of soldiers with a red flag on which appeared in white letters HAIL TO THE NEW GOVERNMENT! They walked in parade-like fashion through the whole village. As they marched, all the other Russian soldiers, except the older soldiers for whom the news was frightening, joined them.

The soldiers, most with rifles with bayonets in hand, chanted socialist catchphrases. "Rise up! Rise up, working people!" That and numerous other slogans were chanted.

The procession ended on the square where the First Battery had assembled. That was the largest group of soldiers.[45] There they assumed long rows. Captain Sukholov went to them and welcomed them. "Hello, comrades!" "Good health to our captain," they replied. "Tsar" was absent from their exchange.

Sukholov—a typical Muscovite with a large, well-kept beard—gave a speech. He knew well the psyche of the Muscovites, so he spoke to them in the manner to which they expected to be spoken. Thus, his speech was received with three cheers.

# The Oath

In a few minutes, Colonel Dudinsky and his staff appeared on horseback. The captain enjoined the soldiers to be quiet and to listen to the colonel. The colonel rode to the middle of the gathering of soldiers. Remaining on his horse, he said, "Congratulations! We have a new government!"

The soldiers again responded with three loud cheers.

Colonel Dudinsky then reminded the soldiers that, new government or not, the war was not over. "Every petty officer or soldier needs to perform his duties because, though earlier we served the tsar, now we serve our homeland, which requires us to be obedient and punctilious in performing our duties. I am the man," continued the colonel, "who will lead you, and I promise to lead you to victory!"

The brief speech was followed by a firing of rifles. The colonel turned his horse and rode with his escort back to his apartment.

The soldiers remained in the square. Some chatted with each other; others shouted out socialist slogans. The First Battery was especially boisterous. Captain Sukholov ordered them to calm down. They did not listen. Yet Sukholov was without his deputy, Adjutant Lieutenant Savcilov. One of the soldiers asked the captain, "Where is Savcilov?"

"At his apartment," replied the captain.

Of the First Battery, five soldiers, each with a bayonet on his rifle, went to Savcilov's apartment. The captain didn't stop them. He turned pale and fell into the cold water of the swamp in front of him. After a few minutes, the five soldiers returned with the body of Savcilov, alive, but fixed in their bayonets.

The soldiers brought Savcilov to the square. He was bleeding profusely, and as they carried him, he frantically kicked his feet as if to free himself from their bayonets. His mouth was open, and there was a deadly wheezing. It was a terrible sight and unpleasant. The soldiers lifted him high enough for everyone to see before they threw him in the icy, dirty swamp, where he soon died.

The captain tried to say something, but no soldiers listened.

A few soldiers took the corpse of Savcilov to dig for him a grave in the wide border of land between the yard and a rural field spotted with various piles. The grave was haphazardly dug. The body was cast into the hole, and he was buried like a dog.

Lieutenant Savcilov was hated by the men. He belonged to those elders who "motivated" his men by doing horrific things to them. While training them, he would strike them and belittle them with inhumane verbal abuse. For instance, only a few days before the announcement of the revolution, Savcilov took a battery he was training through a pond of frigid water. The soldiers quietly cursed Savcilov, though they endured the bullying of the Muscovite. With news of the revolution, he cared nowise to see his men, so they went to him and treated him as he had treated them.

Savcilov's murder inflamed the soldiers, even many of the villagers. Soldiers ignored the orders of their superiors. Each was in command of himself. The result was savagery, and I care not to write more about that.[46] The chaos lasted a week, but in the end, there was relative calm.

When spring came, many farmers, soldiers, and their horses took to the fields for plowing. The colonel had given permission for soldiers to allow their horses to be used by farmers for plowing, so long as he was consulted beforehand. Soldiers who farmed were given food and drink by farmers with sufficient food and drink. Other volunteer soldiers were merely thanked.

The peasants were grateful for such a service because most could not cultivate their fields without help. It was again the initiative of Colonel

# The Oath

Dudinsky, who acted thus much like a foreman. Yet he was careful not to advertise his role. In one of my last visits with him, he told me that the peasants could ask him or other commanders for the help of soldiers to ready their fields. I related the news to the villagers, and the news quickly spread through the village.

By the second day, there were dozens of such requests. The colonel's office kept a record of each applicant and of the appointed day for plowing his fields. When a farmer came to the office, he was given a receipt that he had walked to the location of the battery. There he was appointed a soldier with a good pair of horses, which he kept in his field.

So went the spring of 1917 for the peasants of Khom'yakivka.

# Chapter 16

## Russian Withdrawal

AT THE END OF MAY 1917, The Russian artillery division left us. In its place, there came the infantry, but they stayed only a few days and then moved to the east, where the front had moved. Muscovites were enduring all they could suffer. From afar, we could hear the roar of cannons firing, then ceasing, and then firing again. The cavalry often revisited Khom'yakivka to rest before moving on.

By mid-June, it became clear that the Muscovites were retreating from the Eastern Front. I feared for our vital possessions. After consulting my mother, I took the cart and horses and hid them in a small, dense forest that stretched behind our hayfields. I was joined by seven other villagers—some with their horses and carts, others with their horses. The place where we hid the horses and carts was inaccessible from three sides because it was surrounded by an impassable swamp, and dense vegetation was a thick cover from the fourth side. Tied to trees, the horses would graze by night in the woods and carefully be taken out into the clearing to graze by day. The Muscovites, keeping to the roads, never came near the forest or the clearing.

We remained in the woods for nearly a week and were fed by our brothers and sisters, who would surreptitiously bring us food and drink.

One night, we saw a fire in the village and heard several explosions. Though the fire seemed close to my house, I did not go to the village. Early in the morning, my younger brother Volodymyr came to me and told me that Muscovites had set fire to the barn of Dmytro Zayachyk.

Zayachyk had a warehouse of weapons and ammunition there. The Muscovites simultaneously set fire to the Palace of Prince Lyubomyrsky in the settlement, yet the villagers took many weapons from Zayachyk's burning barn. They had Austrian Steyr rifles and many hand grenades. The rifles and grenades were from an Austrian shop in Stanislavivka. The Muscovites confiscated the weapons and then moved them to Khom'yakivka.

The news of the burning of Zayachyk's barn worried me because the barn was not more than three hundred fifty meters from our barn. Yet my brother assured me that the explosion that destroyed the barn did not threaten our barn. He also said that, in the village, there were no longer any troops, so I could go to the village and stock up on rifles and ammunition. Five of us ran back to the village, and I ordered the three other fellows to guard vigilantly our store in the forest.

Reaching the village, I ran into the house to greet my mother, whom I had not seen in a week. The other four and I then went to a thicket overgrown with thorny plants, where the rifles, ammunition, and hand-grenades were. We took ten guns and as much ammunition as each of us could bring. Walking between fields with the high rye grown tall, one of the guys quietly loaded a rifle and fired in the direction of the village. At that time, there was a Russian Cossack on the road. Hearing a shot that was different than from a Muscovite rifle, he spurred his horse and rode at a gallop to the village of Pshenychnyky. I scolded the fellow who fired the shot and took away his rifle. I forbade the others to fire, because there was the threat that the other Muscovites could burn the village in revenge. That convinced them. They obediently complied with my order.

We reached our refuge. I told the others to hide the rifles and ammunition in a safe and dry place, but that was not so easy to do. The guys with guns did not want to say goodbye so soon to their new "toys," yet I convinced them that holding on to the guns was dangerous. If the Austrians or, even worse, the Magyars had found us with them, they could have hung us. Having appealed to their rationality or sense of conscience, I convinced them to lay down their rifles in the bushes nearby.

I was tired and so I sat down under a tree, where I dozed. As soon as I dozed, the others scurried to the rifles, loaded them, and started shooting. Startled and angry, I soon took away the weapons and, once again, severely rebuked them.

Within a half of an hour, we were approached by the Austrian Cavalry. Circling the marsh, they saw our horses and galloped up to us. Their commander, a lieutenant, said harshly in German, "What are you doing here?"

"Training the horses," I replied.[47]

"Who did the shooting?"

"Nobody here," I lied. "But then, there was an Austrian exploration a while ago, and they fired a few shots."

"Hmm, where is there a village Markivtsi?"

"There." I pointed.

"Where's Khom'yakivka?"

I again pointed to the direction.

"You are fluent in German. Where did you learn it?"

"In the gymnasium."

"Are you from the village?"

"Khom'yakivka," I answered curtly.

# The Oath

"Are there any Muscovites in Khom'yakivka?"

"No. None. The last ran two hours ago."

He was perplexed. "How do *you* know?"

"I was in the village at that time."

"And in Markivtsi?"

"I don't know, but when they are not in Khom'yakivka, it is very likely that they are not in Markivtsi, because they marched at noon to the north and to the east."

The lieutenant looked at his card, which had on it a map, and asked, "Where are we now?"

I came closer and asked for his map. "Here. If you go to Khom'yakivka, please keep to that river. It will lead you to the village."

"Thanks," said affectionately the lieutenant, who spurred his horse and drove in the direction I showed him.

"See?" I told my boys. "See what could happen if they caught you with a gun in your hand?"

The fellows hung their heads and said nothing.

Several divisions of Austrian infantry passed us in the evening. Some were visible; others, we only heard.

At night, the villages of Markivtsi, Khom'yakivka, and Pshenychnyky and the town of Tysmenytsya were occupied by Austrian soldiers who had come from the front at the east. The Muscovites, in retreat, offered almost no resistance, and so the Austrians quickly followed them.

In a few weeks, everything returned to normal, or nearly so. Trains returned to their usual schedules. We began again to get mail. We received news from my father that he was alive and well. The news made us ecstatic.

At the end of July, there was a new Russian mobilization, which took more men, especially the young, from villages and cities. I, too, was called up, but I got a dismissal because my father was in the war, and I had a sick mother. I worked on the farm, and I studied. Good Father Durdello much helped me.

The revolution placed in question Russia's role in World War I. In July 1917, the socialist Muscovites were challenged by Vladimir Lenin's Bolsheviks. On November 7, 1917, the Bolsheviks ousted Kerensky's Provisional Government and tried, under the leadership of Lenin, to steady the situation in Russia. Lenin campaigned to end the war and return home Russian soldiers. With the Treaty of Brest-Litovsk in March 1918, Russia withdrew from the war.[48]

# Chapter 17

## Ukrainian Independence

**Picture 17.1. Ukraine's Push for Independence (1917)**

THE CENTRAL RADA⁴⁹ (ЦЕНТРАЛЬНА РАДА) was founded in Kiev, Ukraine, on March 17, 1917, prior to Russia's withdrawal from the war. The Central Rada, with Mykhailo Hrushevsky chosen as chair, was the first act of an autonomous Ukraine, the Ukrainian People's Republic, and that gave Ukrainians great hope for the future.

We students were happy for the Ukrainian government, if only a provisional government, but older Ukrainians chided Hrushevsky for not proclaiming Ukrainian statehood, because they believed that the time was ripe. The first slogans of the Central Rada, although poetic and sublime, were still formulaic, and all noted that Ukraine was still dependent on Moscow. However, people like Dr. E. Olisnytsky, I. Makuh, B. Yanovich, I. Hryshkevich, Father S. Levitsky, and Father M. Durdello, as well as many others, pushed for Ukrainian independence. They did not share the views of the Ukrainian National Congress, which took place on April 19, 1917, and proclaimed merely Ukraine's desire for autonomy—not its independence and statehood.

My heart was with the radical reformists, though I was too young to have much in common with them. Father Durdello, whose excitement moved me, did what he could to push for Ukrainian independence and statehood, whether the move was smart or not. The Ukrainian skeleton, he said of the members of the Rada, allowed itself to be fleshed out on Moscow meat. Still, Galician Ukrainians were happy about the turn of events. They envisaged a large Ukraine and believed that circumstances would change. Ukrainians merely had to be patient.

Soon the Russian Bolsheviks, under Lenin, turned against Ukraine. They set up another Ukrainian republic in the eastern city of Kharkiv, which was at first also called the Ukrainian People's Republic. The Bolsheviks then led an attack on Kiev on February 9, 1918. The government evacuated the city and fled to Zhytomyr.[50]

News of the campaign of the Bolsheviks in Kiev and the evacuation of the Ukrainian government stunned the Galician Ukrainians. They circulated various unvetted rumors. Yet Ukraine was not fleeing the Bolsheviks, but was waging war with them, under the direction of Ukrainian socialists. Facing defeat, Ukraine, on the day of the Bolshevik invasion of Kiev, turned to its former enemy, the Central Powers of Germany and Austria-Hungary, and signed a peace agreement. The Central Powers drove out the Bolsheviks, and Ukraine was recognized as an autonomous state.[51]

That again had a positive influence on Galician Ukrainians. Though much was not known of the content of the contract, Galician Ukrainians were still happy that Ukraine had been recognized and that the war between Austria-Hungary and Ukraine had formally ended. The Ukrainians in Galicia as well as in Bukovina joyously embraced the Brest-Litovsk Peace, and in Lviv, a large Ukrainian celebration was held. Poland, however, reacted critically, hostilely. They called the Brest-Litovsk Peace the "Fourth Analysis of Poland."

In the spring of 1918 there was a new military conscription. I was called and appointed first to the Eighth Cavalry Regiment, which unfortunately had no horses, and later to the Eighteenth Regiment of Fusiliers.

After training I was transferred to the 117th Unit of the Border Police, which was on duty at Lublin-Lubart. The temperature of Ukrainian nationalism was at the boiling point. Although the Germans and Austrians drove out the Bolsheviks, they could not help the Ukrainians to build their own state.

We received an accurate account of significant events in Ukraine, yet the news was always tardy. In early May, we learned about the uprising of April 29 and that the government in Ukraine appointed as hetman Pavlo Skoropadsky[52]—descendant of Hetman Ivan Skoropadsky from the eighteenth century. That appointment was not welcome news to Galician socialists, but a great proportion of Ukrainians, especially the conservative majority, were happy and rejoiced. There would be some order in Ukraine, and perhaps soon there would be a large Ukrainian state.

In Galicia, there had been very little bread. That changed. Our petty

## The Oath

bourgeoisie and the peasantry produced much bread for all of Ukraine. Without any obstacles, the Austrian authorities were given a pass in Gusyatin, Podvolochisk, and other crossings of the Galicians, and they bought all kinds of food—especially flour and cereals—much needed for their military. There was now some place for every Ukrainian to buy bread for his hungry family.

The hetman's government put up no obstacles for buying bread. It even offered advice to the penurious peasants concerning where to go to buy the cheapest bread—usually in the border towns. The abundance of bread and other products in Ukraine was due to the native intelligence of Ukrainians and to smart economy.

"You see," said some farmers who traveled to Ukraine, "when the government does not abuse its power,[53] there is no hunger. Not only do Ukrainians have plenty of bread, but they are also willing to sell some to us!"

On October 22, 1918, I received a ten-day vacation and came home. My family was happy to see me, especially because it was necessary to gather a meager harvest from our field for the fall. The yield was small vegetables because the late spring and summer were dry and autumn was rainy and cold, and because there were few to work the fields.

On October 31, I had to return to my policing duties, though there were certain various signs that I would not be needed.

First, Austria every day decreased its activities in Ukraine.

Second, there was the Big County People's Assembly, which was to take place October 27, a Sunday, in Tovmach. Father Durdello gave a sermon in the church and encouraged members to go to Tovmach for that meeting of great national importance.

From our village, Father Durdello and his family; Lieutenant Yevgeni

Yavorivsky, who was also on vacation and was visiting his uncle; Ivan Zaychuk, who borrowed a saddleless horse; and I, with horse with saddle, traveled. The Durdello family took a Ukrainian flag, which was decorated on top with tassels in autumn colors.

The day of our trip was rainy and cold. Yavorivsky, Zaychuk, and I rode ahead of the cart in the uniforms of the Austrian army. Much inflamed by Ukrainian patriotism, we wanted to travel quickly, but we could not leave behind the Durdellos. On a good road, their horses could go faster, but on a dusty and rocky road, they could barely walk.

In an hour and a half, we drove to the Tovmach, where there were already several thousand people in the market. The main streets were occupied by Ukrainian Cossacks. They entered the market in couples. In all, there were twelve hundred sabers. Two hundred Cossacks, dressed in black coats and tall black hats and each with a saber, stunningly represented the village Palagychi. The rest of the people were variously dressed—with Cossack hats, Austrian uniforms, and so on.

We joined the people of the village Striganets. From the balcony on many houses, several different patriots, especially Ukrainian Catholic priests, were giving speeches. Others, too, stood out. There were the strong and grandiloquent voices and impressive figures of Father S. Levitsky, priest of the village Striganets, and our own Father Durdello.

About a hundred soldiers formed the guard of honor near one house, and around them, there was a huge circle of people. The speakers recalled the outstanding dates in our history and called attention to significant approaching events for which the people needed to ready themselves. The most eloquent speakers were much applauded, and among them, Father Levitsky and Father Durdello.

It is unfortunate that there were no cameras to film the great event to preserve that historic moment. It was an event so large—something extraordinary, something magisterial—which Tovmach had never seen and never again will see.

Rain fell later in the day and continued through the night, but that did not dampen the enthusiasm of the participants of the unrest. It was not a political gathering. It was a political revolution by the people of Tovmach County. It showed that they cared about their future and the future of their offspring.

Completely drenched, we returned home later that night. Still, we were cognizant that we had participated in such an enormous event in Ukraine's history.

# Chapter 18

## Ukrainian Rebirth

ON NOVEMBER 1, 1918—It was a Friday—I was woken from a sound sleep in the morning by our *Wachtmeister*, Mykola Kotsky, from Tysmenytsya. He instructed me to drive to the villages of Pshenychnyky, Nadorozhna, Klubovets,[54] and Ostrynya as a courier. I was to notify them that Austria had collapsed and that the Ukrainian National Rada in Lviv had seized power and that the District Team in Stanislavivka encouraged all villages and communities to conduct business as usual until further notice.

I received this message in a written form, signed by someone whose signature was fuzzy and impossible to read. Along with it, I got a piece of paper on which the headmen of those villages had to sign off to show that they had received my message.

I quickly got ready—I ate breakfast standing up—and saddled a horse and rode up to Pshenychnyky. It was a mild day for the first of November. Through the autumn clouds, the sun came out. It was nice to pass through the fields where the villagers, digging up potatoes and picking corn, worked.

In half an hour, I was in Pshenychnyky. I asked where I could find the village's headman and was led to him. He was about to go into a field.

"Mr. Headman, I am a courier of the District Team, and I bring joyful news!"

# The Oath

He interrupted me with a gesture.

"So you already know?" I said.

"I found out just before you told me."

"Do you know the order?"

"What order?"

"The County Team encourages you and your villagers to keep the peace and order till further notice."

"Well, thank you. I didn't know that, but I'll try to be orderly!"

"Please sign this paper and stamp it for me."

The headman went inside the house. A minute later he returned and gave me my paper, signed and stamped. "You are in a hurry, right?"

"Yes, I have three more villages to visit. Goodbye."

"Be healthy!"

Next, I rode to Nadorozhna, where it was harder to find the headman. The village was decked out with handsome gardens, and only few people were on the street. Still, with persistence, I managed to find him. This headman was not only ignorant of current events in Ukraine but also seemed not to know anything that was happening in the world. He made big eyes when I told him about the purpose of my visit. After some time, he said, "So, you say there is no Austria—no more?"

"Yes. Now we're the blacksmiths of our own destiny."

"Well, how will things be?"

"It will be just as we will do," I, surprised at his mental sluggishness, replied. "You, Mr. Headman, call together the villagers and tell them what I told you, though please do not forget that the village needs to conduct its business as if nothing has happened. Please sign the paper and stamp."

"But—but why?"

"I need to prove that you've been informed."

"Oh, I see—I see."

While he signed the paper, I asked him whether any Austrian soldiers were in the village. He said none were. I took from him the signed paper and said, "Now, good health and work for Ukraine!"

"May God guide us," the headman responded lifelessly, with a strange blank, emotionless look.

I left his house, got on my horse, and rode toward Klubovets.

The headman of Klubovets was rich and highly educated. He met me with open arms and invited me in and treated me to bread with cheese and sour cream.[55] The news I brought him, he took with great joy and promised to do everything in his power to help in the structure of the fledgling Ukrainian state. His joy was infectious. My young heart was now filled with hope—hope and happiness. We cordially said goodbye, and I began the ride to Ostrynya.

Unaccustomed to such travel, my horse was tired, so I rode at a slowed pace.

It was noon when I arrived in the village, and the villagers already knew the news from the Tovmach. They had just created a provisional militia of some four or five people. The headman met me, sincerely thanked me

# The Oath

for the good news, signed my paper, and said he would make every effort to preserve order and peace.

I asked him if he could give my horse something to eat because I still had fourteen kilometers of road to travel. He gladly did, and in the meantime, he invited me to a small repast that consisted of bread with honey and a cup of warm milk.

About two o'clock in the afternoon I arrived in Tysmenytsya, where I had been given the orders. Here, I followed my orders and handed over my report to the proper authorities. Deputy Commandant M. Kotsky took the report. He then thanked me for the assignment and reported that the Department of Riflemen was sent to our village to disarm and transform the so-called *Wirtschaft* to the Estate Kosiv.

"Will they wait for me?" I said.

"No. Go to your house and rest if you are tired. But if not so tired, I'd be glad if you'd help them in that task."

I said goodbye to Wachtmeister Kotsky and began the trip home. My horse, sensing that she was going home, hurried without a hitch.

When I arrived at my house, I dismounted my horse, gave her[56] water and food, and entered. Mother and my siblings looked at me with amazement when I told them about my adventure.

Evening was approaching. I arrived to help members of the Department of Riflemen, now in my village, for the disarming[57] of the Estate Kosiv. There was there already a guard, from our peasants, led by teacher Volodymyr Mikhailovsky and Petro Kalinchuk. The army arranged for the Austrian soldiers—consisting of elderly Magyars, Slavonians, and Croats—to leave the houses *(Wirtschaften)*. All weapons were placed in one of the rooms of each house. Eight carbines, each with five rounds, were distributed by the watchman, who was replaced every two hours. Each wirtschaft[58] was well managed. It had thirty couples of horses, fifty vans, fifteen heavy plows, eight iron bristles, one hundred twenty seats, and twenty packets of bullets[59] as well as flour, sugar, coffee and even ready-made bread.

When the soldiers learned that they could go to the houses and when their German commandant drove off somewhere unknown, then each soldier did so. The Household Department was overseen by a good mistress under the leadership of a German agronomist, so it had large stocks of hay, potatoes, rye and wheat, and even sprats. All these were inventoried, and one copy was handed over to the District Team in Stanislavivka, and the second one was given to Volodymyr Mikhailovsky. In three weeks or longer, all of this was transmitted to the District Team.

On the second of November, there was a demonstration at Tysmenytsya with the aim of calming the big village Lyads'ke-Shlyakhotskoye, which did not want to recognize the Ukrainian authorities. About sixty people, each with a carbine and bayonet, gathered near the reading room (просвіта[60]) on the square. (It should be noted that Khom'yakivka was the most obedient and well-supplied village. It had carbines, bayonets, and even uniforms of Austrian soldiers.)

Around ten o'clock in the morning, Mikhailovsky, P. Kalinchuk, and I headed from Khom'yakivka toward the village of Markivtsi, three kilometers southwest of Khom'yakivka, where we had to stop to meet up with soldiers Markivtsi and then Odai and Slobidka. The aim was to march to Lyads'ke-Shlyakhotskoye. We waited a long time for those who had to join us on the square and on the road near the Markivtsi church. Somewhere around the first hour of the afternoon, everybody had gathered. We then marched to Odai and Slobidka, where soldiers from those villages joined forces with us and those of Markivtsi. All of us were under the leadership of the wachtmeister of the cavalry, Eustaphy Grobelko. There were in all some one hundred twenty soldiers with rifles and bayonets.

# The Oath

Having traveled two kilometers, we reached the railroad road, Chernivtsi-Stanislavivka, and rain began to fall. Just then, we met and joined a procession led by men with a cross and banners, under the command of the commandant of the gendarmerie in Lyatskim-Shlyakhotsky, a Ukrainian by origin, but a great Austrian. Our commandants cordially greeted them, and we proceeded to Lyads'ke-Shlyakhotskoye.

The march went smoothly, though there was a lack of order. First in order were men carrying a cross and a number of banners, followed by a hundred or so peasants, and behind them our troop of some one hundred twenty soldiers. The commandant had only a revolver and a small police saber. So disorderly was the march that it was hard to guess who was leading it.

Lyads'ke-Shlyakhotskoye was diverse, almost cosmopolitan. There were former gentry Ukrainians there as well as gentry Poles. The peasants, too, were divided into two: Ukrainians and Muscovites. It had two churches. The Catholic Church had three reading rooms: one Ukrainian, one Russian, and one Polish.

We were on a spacious square near the Ukrainian reading room (читальня), where a mass of people was waiting for us in the rain. It was obvious that the people of Lyads'ke-Shlyakhotskoye knew about our accessory, because refreshments were prepared in the reading room, and a platform was built on the square. Yet whether they wished to see just so many armed people, it was hard to guess. When we stood in two rows near the rostrum, we were greeted with the warm words of the old Ukrainian Catholic priest, who expressed his joy that he had lived long enough to see a Ukrainian army and finished with the holy words "O, Lord, help to rid us of our slavery!"

After him, the commandant of the post office gave a speech. He stated solemnly that he was happy that he could now serve his native Ukraine.

Next on the rostrum was the teacher Volodymyr Mikhailovsky. He briefly narrated our story.

After him, Vasyl Vasylyshyn from Markivtsi, a railroad official, moved the crowd with stirring slogans.

Then Mikhailovsky called on me to speak. I had never before spoken and the large gathering was frightening, and so I tried to disappear, but those around me pushed me to the rostrum.

When I went to the rostrum, it was drizzling and cold, and I got a tremor, but somehow I composed myself and said:

> Dear colleagues, although the world is deaf and it is raining, we must feel spring in our hearts—the spring that brought us freedom after five hundred thirty-seven years! Let us rejoice in our hearts, as we realize that we are a free people who serve not others, but ourselves! We shall now work not for someone else, but for ourselves, yet we shall share the fruits of our labors with our countrymen. God gave us the will to be free, and we must respect that God-given gift. Let's all stand as one in defense of freedom! Let's prove that we are worthy descendants of our glorious ancestors! Do we not live on the same soil as did Prince Roman, King Danylo, and Prince Yaroslav? Let us devote ourselves and all our full energy to the service of our motherland, and let's be faithful to her until death! May God help us in this!

That was my first public speech.

There was abundant and long-lasting applause, but I noticed little of it. There was a loud buzz in my head, and it seemed that my body was aflame. Still, I managed to leave the rostrum and get back to my place, though I thought that I might faint.

# The Oath

Many people congratulated me for my speech. The first was Volodymyr Mikhailovsky. I handled valiantly the congratulations.

The old Ukrainian priest closed the series of speeches. In his speech, he, too, congratulated me and praised me for such cleverness from such a young speaker.

Afterward, we were invited to have a snack and tea or coffee. Although it was almost evening, I could not eat much. I drank only a cup of tea and ate an apple tart and a bit of rye bread.

At twilight, we left hospitable Lyads'ke-Shlyakhotskoye. The villagers sadly bid us goodbye. It was a unique day. Together, we had experienced the day of Ukrainian rebirth.

With songs on our lips, we began our trip home. The flickers of light from the setting sun afar seemed to dance on a swamp we passed.[61] It was still drizzling—it had been drizzling all day—but I did not much notice it.

When we reached Markivtsi, we bade goodbye to our comrades of the day. I then rode another four kilometers to Khom'yakivka.

Once home, I threw off my wet clothes, drank a glass of warm milk, and went to bed.

## Chapter 19

## "I shall faithfully serve my motherland, Ukraine..."

ON SUNDAY, NOVEMBER 3, 1918, all who wished to be a Ukrainian soldier had to appear in Tysmenytsya to swear allegiance to the Ukrainian state. Father Durdello, during the Divine Service, sincerely urged everyone who was able to carry a rifle to make the trip.

Again, about seventy people armed with the rifles and bayonets gathered on the square in Khom'yakivka. Under the leadership of Mikhailovsky, the Khom'yakivka Hundred, as we were tongue-in-cheek nicknamed the day prior, cheerfully began the march to Tysmenytsya, each soldier with a song on his lips.

The trip was uneventful, but there was no rain and the road was a little sloped, and we would travel the six-kilometer road in less than an hour.

There was already a large gathering of people in the market at Tysmenytsya when we met in front of the Ukrainian People's House. There were also many other soldiers there, and there were also about a thousand people in civilian clothes who lauded us with their plentiful applause.

The celebration would begin at three o'clock in the afternoon. We arrived some fifteen minutes before two o'clock. Our army formed two rows in front of the main entrance of the People's House, and from balconies, there hung Ukrainian yellow-and-blue flags. At the neighboring "Bank" and Brother's House, flags were fluttering too. The whole city was under

the spell of the great day—a holiday in the making. The clouds began to clear, and rays of sunlight sparkled on the bayonets of the Ukrainian army.

Having a little time, I went to see the Big Hall of the People's House. On the stage, there was a long table covered with white lace. On the table, there was a great crucifix, which partly blocked the Gospels behind it. On the sides of the hall, there were golden candlesticks, each of which held five candles. Everywhere there were Ukrainian flags. In the glow of light, the large hall took away my breath. Never before had I seen anything like this.

From the residence and under the direction of Father Kotsyba, there came into the hall a number of priests, among them Father Levitsky and Father Durdello. Once the priests entered the hall, then we were asked to enter. We entered the hall and took the first rows of chairs before the stage. Candles were lit on the table of the stage, and the priests stood behind the table. Old Father Kotsyba began with warm words and a prayer concerning "The King of Heaven." Afterward, Father Levitsky, the pastor of the village of Striganets, gave a moving, well-prepared speech. He was a wonderful speaker and a spirited and loyal patriot. In his speech, he urged all of us, without reservation, to serve Ukraine, and if necessary, even to give our lives for the cause of freedom. He spoke briefly, but each of his words sunk deeply into the soul of every listener. Father Levitsky was so captivating, so persuasive, because he spoke not only to the heart but also to the head. So much did everyone wish to hear every word he uttered that the room was absolutely quiet while the patriot-priest spoke. So quiet it was that it seemed as if he were speaking to an empty room.

Having finished his brief speech, the priest took in his hand the Gospels and asked everyone to recite an oath to Ukraine, which, to the best of my faulty recollection, went something like this:

> I swear by the Lord God in the Holy Trinity, before the Holy

Cross and before the Holy Gospels, that from now till the end of my life, I will faithfully serve my motherland, Ukraine. I will honestly and faithfully perform all the duties imposed upon me, and if required to do so, I will readily give up my life for Ukraine, so help me, God.

**Picture 19.1. Chemny in the Uniform of an Otaman Sich (1926)**

The words of the oath also sunk deeply in the souls of my five hundred comrades in this great celebration. It was an unforgettable moment for everyone.

More, however, had happened to me. I was no longer standing on the ground, but I was somewhere in Heaven and was the happiest of all people in the world. At that moment, I fully experienced my whole being. It was something that I cannot now express in words—perhaps something that cannot be captured by words. I can say that it was a transformative moment, the happiest moment in my life. Never before had I experienced such bliss, such mental well-being, and I cannot expect I shall ever again experience such a blissful moment. With that oath, my world widened—eternity was, in a sense, captured in a flicker of time—and to this day, I mumble in my heart and head the gist of the oath and shall take it to my grave.

I later had to swear an oath in the UHA ranks, yet I did not feel the same joy. Even today, with my advance of age and in a foreign country, I plainly see the larger-than-life image of Father Severyn Levitsky with the Gospels in his hands and five hundred Ukrainian soldiers, each of whom, in making an oath, gripped tightly a Ukrainian rifle in their hands.

About a week after the celebration, my father returned home, and for a long time, my family rejoiced. My father warmly greeted me but did not share my desire for me to go to the front. We argued often. He said that the front was the front, not the rear, where it was safer.

In the end, I still did not listen and I went to Stanislavivka and became part of the First Striletsk Regiment. I was soon assigned to the Second Hundred Division, and the enemy was now Poland.

On November 18, I was stationed at Stare Selo, some twenty kilometers southeast of Lviv, which at the time was predominantly Polish. On November 21, the Ukrainian troops in Lviv were driven from the city. Thus, our struggle for life and death with the Poles began.

Our part retreated to Vynnyky, just outside of Lviv and southeast of it. There, during the offensive, I was wounded by a grenade, which injured my foot. I was transported to a hospital in Stanislavivka. Doctor O. Dyky stated that I had a broken bone in my foot,[62] which would take much time to heal properly, and so I could rest freely. My parents, having found out about my injury through our peasants, came the next day to visit me. The hospital was crowded and small, and it could not take in all the wounded and sick. Therefore, my father asked Dr. Dyky to let me go to my house to help ease the overcrowding and where it would be easier for me to recover. The doctor said this was possible, but that he could not issue the order and it would take a few days to move through the order. In addition, he wanted to wash well the wound and to give me a fresh bandage.

I would stay in the hospital till February 1919. With it being overcrowded and with there being so many open wounds, the doctors also had to fight typhus. I, too, did what I could to comfort others in the hospital.[63] So Dr. Dyky told me that he would like to send all who were not too sick to a place with fresh air. For me, the stay was insufferable, diversion was impossible. I could not read or write because of the constant moans and cries of the poor soldiers.

Sometime around February 25, I went home with my parents and younger brother, with the condition that when I recovered I would rejoin the Ukrainian front. My wound, however, took much time to heal. I even washed it with cream and water and with kvass, but my broken bone was so painful that I could not wear shoes until May.

Things on the front with the Poles were bad. The Poles were supported by Galicians, and we had many deserters from villages and cities. There were many expeditionary cargoes sent to catch deserters, who, having left the front on the second or third day, would run from their houses when pursued and hide in the boscage and in the woods. The worst thing was that the deserted fled home in the uniform of a Ukrainian soldier.

Thus, desertion paralyzed our combat capability. That tugged at my soul. Had not those soldiers taken the same oath that I had taken?

## Michael Chemny

In the second half of May, the Polish front advanced to Stanislavivka, some seven kilometers to the northwest of Khom'yakivka. On May 25, Ukrainian troops evacuated Stanislavivka.

It was a Sunday. The morning was pleasant, and in the evening, there was a dense spring rain. In our village, the third-largest Zolochiv regiment arrived and moved eastward. I almost cried when it left our village, because I wanted to join the men. I could not because my left leg was still bad and I did not have the strength to go with them.

I would remain at my house until the arrival of the Poles.

# PART III
# POLISH YOKE

# Chapter 20

## Poles Invade Ukraine

AT ABOUT THE END of June, I was feeling healthier and freer to walk, although my leg was still in pain. The Polish Army was marching from west to east and had no intention of remaining in our village. Yet while passing through Khom'yakivka, they tortured our people. Soldiers raped women, even girls, and they stripped off clothes from and beat men on the street. It seemed to me that that was God's way of punishing us for those evil deserters who stole Ukrainian weapons and fled from the front. "It was a repentance from which there could be no return," as our proverb says.

Thus, only a few men really served Ukraine from all the seventy to eighty available men from our village who were called to the army. The rest hid in basements and boscage or deserted the front. When the expeditionary cargoes came for the soldiers, the men, with rifles in hand, fled to and hid in the woods. I was massively disappointed. Our village, though small, was proud, yet in the first quarter of 1919, we had as many as sixty deserters. So much for the Khom'yakivka Hundred.[64]

I must go back a bit in time to explain. In the second half of March, criminal investigators came to our village. In the maidan[65] near the reading room, they flogged the parents of deserters until the deserter showed himself. In this manner, Haydamaks had ferreted out some fifty deserters. The deserters were brought to Stanislavivka, where, on the second or third day, they wore civilians' clothes and were paraded through the streets of the village and other men fired rifles from windows to emphasize their cowardice. It was shameful.

# The Oath

I would like to have omitted this story of deserters from my village, but I cannot. It is disgraceful, and it is why we lost the war with Poland, which thereafter brutally made us pay for our cowardice. Were there no desertion, we would have had an army as large as five hundred thousand, not one hundred thousand, soldiers—that is how much desertion influenced events—enough to ward off the Polish advance. For example, in the battle of Chortkiv, our soldiers made every effort to repel the Poles, but the enemy was twice as large as us at the front. The valiant Ukrainian soldiers, outnumbered at the front, cursed the Poles while being beaten by the butts of rifles in the face. Still, the Ukrainian soldiers fought till the end. Overwhelmed by numbers, they still kicked a foot or threw punches to the ribs of their assailants before dying.

When I, on crutches with the help of my brother, arrived at Easter at the church, the church was filled young people in uniforms, most of whom did not serve their native land even one day. Some were so brazen that they even wore their stars on collars. My soul was rent asunder, but there was nothing I could do. Those people had no sense of patriotism. I could list them here, but I have chosen not to do so. Having grown up with most of them, I knew them well.

The Poles presumed that there were weapons in our village. So in July 1919, they ordered that all weapons must be surrendered. If not surrendered, they would seek out the weapons and severely punish, even kill, anyone found to have a weapon. The public chancery found fifty rifles, a dozen bayonets, and two revolvers surrendered, though there were certainly many more in the houses.

**Picture 20.1. Polish Army Advancing**

In addition to the voluntary return of weapons, Poles carried out "revisions" everywhere. They ransacked the houses and stole eggs, cheese, sour cream, and cornflakes. The peasants' houses in the middle of the village looked like a hurricane had devastated them. The women cried and begged, but it helped nowise. Instead of mercy, they received a blow in the face or an injury from the side of a chair. In the most favorable instance, the women were cursed with vulgar, filthy Polish words. We were severely punished on account of our "forest heroes," who learned the hard way—one can only hope—just who the enemy was and why we were fighting the Poles.

The atrocities continued until late autumn, when things quieted and life in the village resumed. There were a few weddings among the young—there was even held a noisy wedding—but at each wedding there were two or three Polish policemen, whom nobody invited, who watched over the festivities, and who insisted on having their fill of drink.

Things became quieter in the winter, but then arrests began in January 1920. Among those arrested, there were Father Durdello and later Volodymyr Mikhailovsky.

# The Oath

It was obvious that all these arrests occurred because of denunciations of Poles from our people. Father Durdello was released after interrogation. Mikhailovsky was released after a week in jail, and he fled from the village. I, too, was interrogated by the Polish commandant of the post office, Stoyalovski, a former friend of my father through military service. After a few questions, Stoyalovski said that I was free to return home.

In all, Polish political leaders, unlike Polish leaders in Galicia, used arrests indiscriminately in Ukraine. It was generally sufficient to justify an arrest by blood—that is, one being Ukrainian and not Polish.

Early in 1920, many of the farms that burnt back in 1915 had been mostly rebuilt. Things began to return to normal in the province, but we still felt the atrocities of the state of war. Telegraphic wires were closely guarded, and there was a Polish police station, centrally located in the village, run by Polish thugs and gangsters.

# Chapter 21

## "A coffin for me!"

IN ADDITION TO THE MERCILESS persecution of the Poles in the winter of 1919–1920, typhus in varied forms visited hundreds of Ukrainian villages.

Khom'yakivka was not exempted. In our house, the older of my two younger brothers fell ill, and then did my sister. Neither, fortunately, was feverous for more than a few days, and thereafter, neither was confined to bed. On February 8, 1920, the Feast of the Three Saints, I returned from church atremble and with a violent headache. "It seems that it is now my turn," I said to my family.

Not waiting for the evening to recline, I went early to my bed. I was cold, then hot, and then cold and hot again; and once, it seemed that I was hovering above the house and then somehow beneath it. On the second day, I was so ill that I could not leave my bed. I was thirsty and asked for water, but I could eat nothing. In the evening, my fever worsened.

Having the fever much worse than my brother and sister, I remained in bed for more than a week. Dr. Yanovich finally arrived. He had been busy with other cases in the village, with two or three people, both old and young, dying each day. He prescribed certain medicines and ordered me to keep clean and warm. I was horrified by the doctor—this my family later told me—and tossing and turning, I was very restless on the bed before I finally fell asleep. When I awakened, I again continually tossed and turned.

# The Oath

A week later, Dr. Yanovich returned and said that my crisis had passed but that I would still be very weak for three to four weeks, and therefore, I needed to be cautious lest my fever returned. I needed also to remain in bed every day.

I asked Dr. Yanovich about my lack of consciousness, which I found embarrassing.

He smiled and said, "That's nothing! It has happened to many—many of my patients with the fever—even those healthy, those very healthy."

I remember little of those feverish days, and the fever lasted longer because we were not equipped to fight it. Each morning, my father and his brother would change my bedding and switch my mattress to keep me clean, and I came to expect that as part of the daily routine.

One day, when I still had a fairly high fever and I was still very weak, my father did not come to change my bedding. Panicked and delirious, I called out for him but was told that he went to Stanislavivka, several kilometers south of Khom'yakivka.

"Ah, this *father* went to find a coffin for me!" I said with reproach and anger. Thinking I was going to die, I began to cry, and my mother and siblings began also to cry. All day, I was told, I called out to my father, and I did so even in my dreams.

In the evening, my fever worsened. I fell asleep, but my breathing, I was later told, was hard and irregular. My mother, thinking that I was moribund, cried loudly and rubbed my body as if to invigorate my vital spirits. Her sobbing awakened me, and I again called for my father. I was told that he was on his return from Stanislavivka, but he was still far away and the road was muddy, so it was impossible to hurry. Mother gave me tea with citron and calmed me so that I again fell into sleep.

**Picture 21.1. Chemny's House Today (photo courtesy Stephen Fedak).** With the passing of Chemny's sister, Efrosina, the Chemny house changed hands and it is now in a state of considerable decay.

Unbeknown to me, My father did not go to Stanislavivka to get me a coffin but to get boards to make coffins for those many persons who succumbed to typhus. All the ready-made coffins were taken.

When my father arrived in the evening, he ran to the house without unhitching the horses. I had just awakened prior to him entering the house.

"Daddy!" I shouted with gladness. I tried to sit up in bed, but I fell back on the pillow again.

My father ran to me, held my head in his hands, and placed a warm kiss on my forehead. Tears fell from his eyes. I grabbed one of my father's hands, and with all my strength, I pressed it to my mouth and kissed it. The rest of the family was moved to tears.

I looked into my father's eyes and, with a weak voice, asked, "Did you bring the boards for my coffin?"

"No, son, I did not bring boards, but certain other things."

"I know about the boards," I said.

My father looked around the room and said, "Who told you this?"

"Nobody, but I saw you, as you—as you—you placed the boards in your carriage in Stanislav."

"How could you see me when you lay here?"

"A dream. I saw you in a dream."

"No, son, it was not a dream. It was your imagination. You have a very hot fever and—"

I interrupted. "Is it true that you bought six boards?" I got from my bed and stood up.

"Yes, it's true, but those boards are not for you! Don't worry. You'll—you'll live!" With those words, my father placed a cold hand on my forehead. I welcomed the coldness. I returned to my bed and looked into my father's eyes, still full of tears.

"Anna," my father said as he turned to my mother, "there in the bag—there are oranges. Give him some—some juice to drink."

The juice was cold and had a pleasant smell. I gulped down the cup of juice and immediately felt invigorated. "Thank you," I said, as I fell down on my pillow, wet with sweat.

"He called for you all day," my mother, in a drained voice, said to my father. "It was such a restless day for him!"

My father clutched my hand and then stroked my head. Lost in thought, he then turned away from me.

In the middle of Great Lent, I could finally get from my bed, but I was still very weak. My hunger returned. My good mother, who was very devout and was fasting for Lent, fed me milk and eggs and fresh, delicious bread. Thus my health was slowly restored.

No sooner than I regained my health my nurturing mother got sick, and in three days, my dear father, too. Both were tormented with fever, and their suffering made the rest of the family suffer. My father tossed and turned, as did I when sick, and was bothered by the least disturbance. My mother lay calm in bed. She looked at us with longsuffering eyes, and she whispered prayers as she lay. Their illnesses were big blows for our family. The loss of either would have been crippling; the loss of both, shattering.

My illness had lessened, and so I, weak as I was, gathered myself and began the long ride south to Stanislavivka to see Dr. Yanovich. He was not in his office, and I had to wait for his return. Those moments in wait were the longest of my life.

I brought back with me Dr. Yanovich.[66] He examined my parents and others in my extended family who were sick, prescribed medicines to be purchased, and gave us instructions on how to convalesce the ill. He then asked me to take him back to Stanislavivka.

**Picture 21.2. Efrosina Chemny (photo courtesy Stephen Fedak).**

Not fully recovered, I was tired of traveling, so I bid my brother to take back the doctor. I gave him money for medicines to be picked up in Stanislavivka and reminded him not to be too hard on the horses, because they had already done thirty kilometers. My brother hitched the horses to the cart, adjusted the seat for the doctor, and the two departed for Stanislavivka. I was left with my sick parents.

My sister, Efrosina, greatly helped me to heal our parents, although she was only fourteen years old. We made cold compresses, cleaned them, and fed them. There often were days that neither of us got more than a minute of sleep.

Easter was nearing, and there was an abundance of work to do to prepare the disordered house and yard for the great holiday. Yet there was no one to do that work because our entire focus had been on helping our sick relatives.

Yet by the week of Easter, our parents were well enough to leave bed and sit, though they were not strong enough to sit for long. My sister went to our aunts to ask for help to whiten the house. Three came. We asked an uncle to whitewash the walls of the house. We told him that the lime would freshen the house and kill a lot of typhoid bacteria and that, consequently, our father and mother would feel better. For half a day, several young girls and newlywed women finished whitewashing the courtyard. We asked our uncle to whitewash the two walls and the stove. Hating lime scum, he immediately refused but eventually was persuaded to do so. In two hours, the house was whitened and smelled of freshness. We then gave our sick parents fresh bedding. They instantly became more cheerful.

Easter of that year was not very joyous for us. My parents were still bedridden. Yet my mother's sister, Luba, came to our house on Holy Thursday and brought us honey. She also brought us much comfort and happiness. I bought meat from the butcher shop. Thus, with the help of relatives still living by the grace of God, our dinner table that Easter was not empty, and though not happy, we had no reason to be sad.

Still, many of our fellow villagers had anything but a joyous Easter. Many lost a father, mother, sister, or brother to that terrible typhus.

# Chapter 22

## A Week in "Dibrova"

ON APRIL 19, 1920, I was again arrested for no reason and severely beaten by the butts of the rifles of Polish soldiers, which broke two ribs on the left side of my chest. In this condition, I was brought to Tysmenytsya, and from there to a jail in Stanislavivka. Stoyalovski, the deputy county police commissioner of Tovmach, then was not there. My father—who at that time was also out of the house—knowing about my fate, now wrote a letter of protest to officials in Tovmach, and tried to free me with Stoyalovski's help. It was said that I was arrested for anti-government activities.

I was interrogated often, but they learned nothing from me, mostly because I had nothing to say. I did nothing against the law.

I felt great pain in my chest, especially when breathing deeply. In four days, a certain Jewish doctor visited me. Bidding me to lie patiently, he stitched cuts in my chest. He did not tell me that I had broken ribs.

After spending almost a week in Dibrova, the name of the prison in Stanislavivka, I was finally released on account of the intervention of Stoyalovski. I immediately sought out Dr. Yanovich and asked him to examine my chest. While he examined me, I told him about my arrest and beatings. After examining me, Dr. Yanovich asserted that I had two fractured ribs but said that I should not panic, because they would soon heal on their own. However, he advised me to rest more and avoid all sorts of strenuous work.

# The Oath

After seeing the doctor, I went to see Haim Vaks, a painter and former resident of Khom'yakivka, who had moved to Stanislavivka. It was difficult for me to speak at his house, in the presence of his father, so I asked Vaks if we could go to the market and ask some merchant to help me. He found Vasyl Yasnykovsky, who lived in Khom'yakivka, and asked him to take me to my house. Vasyl happily agreed and took me to Khom'yakivka.

Late in the evening, I got home. Father, melancholy, was sitting at the table, but seeing me now home much gladdened him. My mother cooked me a snack and warmed some milk for me. The ride on the cart had tired me. I ate and retired to my bed. I quickly fell asleep.

In the spring, I felt better but was still in no physical shape for hard work. So I asked my father if I could go attend a seminar conducted by the well-known teacher and Pole Stanislav Grabik in Stanislavivka, if only to have something to do. The seminar would begin at the end of January. He consented.

With the consent of my father, I went to see Father Durdello to get more information about the seminar. Father Durdello advised me to attend but warned that the seminar might not last long because the Poles could intervene, as the war was not yet over. Moreover, there was also the threat of a Bolshevik invasion.

Father Durdello said more. He secretly told me that he would soon leave Khom'yakivka. He wanted to go somewhere in the mountains to get away from the village. He was greatly agitated, as he said the villagers were a hundred times worse than they were before the war. "I will not play the village mistress," he said with frustration, "I haven't strength anymore—no more. I have lost my faith in the goodness of the peasants. So it's better for me to retreat to the mountains. There, the air's cleaner—the work is less. Mykhailo, nothing has turned out as I had hoped. The Lord God continues to punish us for our misdeeds."

I did not know what to say, for what he said about the peasants was bitter, but true.

In a few days, I[67] went to Stanislavivka to inquire about the exploration workshop. The seminar, which was not yet in full swing, was held at Trinitarian Square. There were in attendance a few students, but Director Grabik needed more equipment, as much of the equipment used in prior seminars had been destroyed in the war. The director warmly welcomed me and said that I could certainly attend the workshop as a student.

"But I can't come till next week, Mr. Director. Is that a problem?"

"No. Not at all. Don't panic!"

"Okay. My father will be happy to know I've been accepted, yet—yet he's worried by payment."

"Somehow, we'll make things work," said the director.

# Chapter 23

## A Visit by a Polish Commissar

SOMETIME IN MID-MAY 1920, my father and I—my mother and siblings were working in the fields—were visited by Commissar Tolmach of Starostvo and two policemen. The commissar told my father that he was to accept a governmental appointment as governmental commissar until the upcoming elections.

My father, shocked, made big eyes. He could not believe what he had just heard.

"Well, what does the gentleman say?" the commissar said in Polish.

"I don't understand! How can the government impose such a duty on me without asking? And, you—your police recently arrested and beat my son for no reason!"

"My orders come from the yard. As there are no Poles in your house, you must accept this position. If you refuse, we'll be forced to send some person, a trusted person, obviously a Pole, to live with you. As for your son, we made a fatal mistake, and we apologize. You know the war is not yet over."

"The arrest can be forgiven, but the beating and broken ribs are not easy to forgive! Look at him! Still a young boy, but he cannot move," said my father in a trembling voice.

## The Oath

"Yes, indeed, that was a very unpleasant event, an unfortunate event, but it happened. Please note that the police officer who arrested and hit your son was released from the service."

"He was not merely hit. He was severely beaten, and it will not be easy for my son to forgive," said more sharply my father. "I can't serve a government that has given me such a heavy heart!"

"I understand well," commented the commissar. "I would never have acted as the officer did, but we have to forget what happened."

"No, I can't forget, and—and I can't accept this duty. I have my own affairs, which require a lot of time and a lot of hard work."

"I merely ask the gentleman to think over my proposal. I'll return in two days. I have other villages to visit. I'm merely doing what other officials are doing through the country."

He got up from the chair and said goodbye to my father and me and went out. My father walked him to the gate and reassured him that the barking dog would not harm him. He then returned to the house.

"Huh, even the dog does not like the Pole! And they—they want me to serve them," he said with disgust. "Well, a new trouble! Damn him! He has no right to come to this house!"

My father sat in his chair and thought. He rolled a cigarette and lit it. He was struggling with his thoughts and did not know just what to do.

"Father, don't be angry. We'll—all of us—we'll figure out just what to do," I said.

"You're right. All of us—that—that might be better. Meanwhile, tell no one why the Pole came here."

In the evening, my mother, sister, and brother returned from the fields. My father informed the whole family about the visit of our uninvited guests.

Mother, who was normally quiet and reserved, was angered by the commissioner's order. "If you accept their appointment, then we'll have no peace in the house. Every Pole who is hungry will come to us."[68]

"What do you think, Mykhailo?" said my father.

"Mother's right, but here's another—"

"What's that?" interrupted my father.

"If Father doesn't accept, then the Poles will send us some kind of Mazur[69] that will tear off our last bit of skin. In my opinion, it's easier to serve some stupid official than to have a Mazur scrutinize us each day through the prism of Polish ignorance and rob us of every bit of dignity."

"What do you say, Volodymyr?" said my father to my younger brother.

"I think that Mykhailo tells the truth. If a Mazur comes, not only we but the whole village will suffer much."

"Fryzia?" said my father to his sister.

"I agree with Mykhailo and Volodymyr."

My father did not ask my youngest brother, Yaroslav, as he was only ten years old.

Reflecting briefly on what he had heard, my father came to his decision. "I see that all of you except Mother have the same opinion, and I can't say that it was not smart, but you must consider what such an appointment will mean for me. Everything, as the mother of a vagrant says, will shake me on all sides. It will keep me from my work, as I will often have to ride to Tovmach and spend time in vain."

"That's true, Daddy," said I. "You have spoken sincerely. Yet if you accept the position, you'll give yourself a chance to play with the Poles. In addition, knowing them, they're ready to punish us, given your refusal. They'll repay your refusal with various injustices in an effort to destroy us not only materially but also physically. I think it's better to accept their temporary proposal and all its temporary inconveniences."

"What do you say, Anna, to what our Mykhailo has said?"

"It seems to me that what he says is good and true, but still I object to your appointment, though I wouldn't like to have a Mazur wake me up at night to burden me with his different whims."

"Yes, yes, Mummy," said I. "You, too, are right. But it seems to me that it is easier to be burdened by someone else's whims from the door of the post of government commissar than to endure those whims at home. It's about whether this will turn out good or bad for us. True, it's not now an easy affair, but when one has to choose between two evils, one has to choose the lesser evil. In the end, this is Daddy's business and Daddy himself has to decide. We have two more days. Maybe tomorrow will bring something better."

My father got up from his chair and came up to me. He took my head into his hands, kissed me on my forehead, and said, "Thank you, son. You think rationally. And now let's go to sleep. We had a hard day today. I'll rethink all this tomorrow and make my decision."

We prayed together and went to rest.

My father, rolling from side to side, could not sleep. He got up and smoked one cigarette after another.

He returned to bed late in the morning and then fell into a sleep so deep that he was still asleep when the rest of us woke in the morning. We did not dare to wake him up. When breakfast was ready, he woke up and said with agitation, "Why didn't you wake me?"

"Well, you needed the rest, as you were up for most of the night," I said.

We sat to breakfast. Father, having pondered the proposal of the commissioner, decided to accept the offer because refusal would be worse for us and for the village.

My father's decision brought us comfort, but he then added that he would only accept if certain conditions were met. One of those conditions was that I was to be his clerk.

"And my school?" I said.

"Things will work out somehow."

On the second day, the commissar returned as he had been commanded. My father accepted the post so long as his conditions were met. First, my father asked that the police not make daily rounds and make chaos in the village. No policeman was to come to Khom'yakivka without his consent. When a policeman was needed, my father would call for one. Second, my father requested that the villagers' carts would not be appropriated by Polish officials for their purposes, because the villagers needed their carts for their work in the fields and the state needed bread. Third, the commissar was to establish me as my father's clerk, and he would appoint my father as a deputy with authority over other officials.

Having heard of my father's willingness to accept the appointment and his three conditions, the commissar showed his appreciation by accepting all my father's conditions, with some reservations concerning the first. He then promised to take everything into consideration. The commissar then left with his police.

# The Oath

He went to the village head, who was elected by the Ukrainian authorities, and took his public seal and all the needed books and protocols and brought them to our house. Having placed everything on our table, he took from his bag the official charter letter and handed it to my father. At the same time, he asked my father to sign the two copies of the letter, and I was asked to sign as a witness the two copies, which my father and the commissar signed. The commissar took one copy with him, and the other one remained with us.

Thus, my father became an "obedient" official of the Polish government.

# Chapter 24

## The Polish-Bolshevik War

IN THE EAST, THE War with the BOLSHEVIKS continued. Ukrainians in the Petliuran Army[70] joined forces with the Poles, as the Russian front moved closer to us. Sometimes we, in Khom'yakivka, heard the sound of the guns firing.

After my father's appointment, I returned to Stanislavivka to visit Stanislav Grabik, the director of the seminary. Engrossed with some paper in his small office, he sat composed. He asked me to have a seat and to wait a minute till he finished his business. He soon put down the paper and said, "My young sir, it seems we need to wait a little longer with your introduction to the seminar. We live in anxiety from day to day and are afraid that the front is still going to reach us."

"Is it so bad?"

"It *is* so bad."

We talked for a while—he was being polite—and I said goodbye to the director. He promised to notify me by mail if circumstances changed for the better.

Having settled my affairs in the city, I returned home. In the evening, I went to say goodbye to Father Durdello, who would be leaving Khom'yakivka in two days. With tears in our eyes, we bid each other farewell. The village was losing its pastor, and I was losing a true guardian and kind friend. We would write each other thereafter, but the letters became fewer over time. In my mind, Father Durdello had lost his faith; he had become sick.

Hopeless days drifted at a snail's pace. My father was engaged with his commission, and I wrote various proposals for his office, which almost every day were a whole heap.

Now after the harvest, when people finished with the harvest, the Ukrainian National Republic troops led by Symon Petliura, who had an alliance with the Poles, approached us. Savchenko[71] and some servants ran up to our house and demanded twenty carts. My father replied that he would be happy to try to supply them if he could give him a date by which he needed them, because there were not even five in the village.

"Why?" Savchenko angrily said.

"We merely don't have many. The war destroyed the village."

"I do not see the village in ruins."

"You do not see ruins because you do not know what I'm talking about. I'm not speaking of ruins in the physical sense, but in the economic sense."

"Indeed," Savchenko said with some suspicion in his voice. The soldier turned to me. "You are in the army for the first time?"

"In which army?" said I.

"The Ukrainian army, of course!"

His manner of speaking, a mix of Ukrainian and Russian words, annoyed me. In Central and East Ukraine, people commonly use fifty percent or more Russian words when speaking. This continues to this day.

**Picture 24.1. Polish Light Cavalry (Uhlans) of the Polish-Bolshevik War**

"Ukraine has an army?" I replied with noticeable cynicism. "You know, sir, I'm sorry to say this to you, but I must say this: I'm ready to join the Ukrainian army today, but not under the Polish flag. The Poles will use you as long as they need you. When the need is gone, then they will arrest you."

"Well, what kind of person are you? We have a treaty with the Poles."

"You do not know Poles. They are polite to you when you are needed, but when they no longer need you, they'll shoot you like a dog!"

"Well, how do you know?"

"And what will happen," I continued, "if the Poles form with the Bolsheviks a truce or even an alliance? Where will you find yourself then?"

"Well, do you think that's—that's possible?"

"It's possible."

His head lowered, Savchenko thought something and then said, "Yes, I agree. The situation, then, is bad."

"The situation is bad, but the biggest problem we have is that Ukraine is not yet an independent nation and does not even wish for independence. Ukrainians are emotionless. Ask yourself whether it was necessary to abandon the Hetman state, when we had a golden opportunity to hold it and secure it."

"You are right!"

"You are hungry, sir?" He looked weak.

"Yes, I could go for something to eat."

"Mom, give him something to eat!"

"But I do not have anything except bread with cheese and sour cream."

I looked at Savchenko to see if that would be adequate.

He said, "I have had nothing to eat for two days."

Mother gave him the food and a cup of milk. He ravenously took to the food. When he finished and drank the milk, he became more relaxed and friendlier. I offered him a cigarette.

"You know, sir," I said, "you will not get a cart because we haven't any. There are some five horses with carts in the village, and those are needed to bring crops from the fields. Maybe you'll find carts in another village that is more fortunate than ours." I paused and then added, "Your horse has also been given something to eat—oats and water."

"Praise God!" he said. "You are good people. Goodbye!"

"Be healthy!" I replied.

Thus, we gave nourishment to one Petliuran, who wanted twenty carts where there were not five. Yet each day, one or two more officials arrived, even a whole group of Petliurans, demanding carts.

# Chapter 25

## The Black Division Arrives

ONE NIGHT, IN THE middle of August 1920, I was asleep on some hay in the barn. Late at night, I was aroused by the barking of dogs. I went to our gate, where there were seven or eight men on horseback at the gate.

"What do you need?" I asked in a voice loud enough to be heard over the barking dogs.

"We need rooms," one of them said.

"Well, then, take what you can find."

"But we have a whole division."

"Well, there's no place for a division, but take what you can find," I said, while I made no attempt to hide my irritation.

"Are you the commissar?"

"No, I'm his son."

"Where is he?"

"What do you think? He's doing what others do at this time of night—*sleeping*!"

"Bring him here!" demanded one of the men.

## The Oath

I went to wake my father. He was already on his feet and about to go outside because he, too, was waked by the dogs that continued uninterruptedly barking. My father went to the gate and asked, "What do you need, good people?"

"We need places to rest our soldiers. We have a whole division."

"Well, get a flat, but we don't have a place for a division because the village is small—only two hundred homes."

"Can our colonel settle down?"

"If he must."

These were the soldiers of the so-called Black Division, under the command of Colonel Didchenko. They made some large sign on the gate with chalk and departed.

By morning—it was Saturday—Khom'yakivka was overrun by cavalry. At night, a fat colonel, dressed in black, came to our house and took my bed. His entourage found spots on the floor. Our yard was full of horses that ate oats that the Cossacks of the division brought them.

On Sunday morning, everybody slept until noon. The peasants came to us and complained that the Cossacks were eating all of their bread. My father sent them to the colonel, but the colonel did not do anything about it. "Well, yes—yes indeed! That's just the whole thing. That's how these things—these things go."

Didchenko's Cossacks treated us worse than the Poles.

Somewhere around one o'clock in the afternoon, one Cossack, wet and covered with mud, rode into our yard and, without any warning, went to the colonel. He told the colonel that he had been beaten by the river by some men—that he had been pulled from his horse, struck repeatedly, and thrown into a swamp.

The colonel did not ask the Cossack whether he had done anything to merit the beating. He merely gathered one hundred horsemen and one hundred speed-shooters to punish the perpetrators.

The beaten Cossack did not mention his trip to the village Chornoliztsi, four and a half kilometers southeast of Khom'yakivka—the two villages divided by the Vorona River and its small Rokytna pipeline. There, he had raped a young woman and robbed a house.

The Cossack could not find the spot where he was pulled from his horse and beaten, so he and some two hundred other men rode to Chornoliztsi, where they took their revenge on the villagers. There, they beat up many peasants and lit fire to several farms.

Returning from Chornoliztsi, they espied on a large lawn a sizeable herd of cattle in the Rokytna River and a gathering of boys and girls with their mothers and some elderly women—all from Khom'yakivka and Markivtsi. The Cossacks rushed at them with naked sabers and began striking them. Speed-shooters then formed a semicircle around those still standing. Little children, insofar as they could, hid in the potato plants. The elderly, overwhelmed, merely stood and waited to be slaughtered. In the end, twenty-two young people and one girl were killed or wounded and one small boy was wounded in the foot. All of them were brought to us on the square near our stall. All were all bloody, and it was frightening to look at the innocent victims.

The colonel, my father, my whole family, and many Cossacks came to the square. Looking at those poor, innocent crippled people, my father asked the colonel anxiously, "Mr. Colonel, what does this mean? What are we to do with these poor, wounded people?"

"They beat our Cossack," he said curtly.[72]

"Do you have proof that they beat him?"

"Yes," he said.

"That's not enough, Mr. Colonel!" angrily replied my father. "Why don't you ask the 'poor, beaten soldier' where he's been—what he's done?"

"Okay, let's do so." He ordered the Cossack to come to him. The Cossack did so.

"Who beat you?"

"They did!" and he pointed to the mutilated women and children.

"Where did they beat you?"

"There," he pointed. "Over the river."

I asked the colonel to ask the Cossack to point out which river, because there are two rivers between Khom'yakivka and Chornoliztsi, and the rivers are three kilometers separated from each other.

"What river?" said the colonel.

"The one above—the larger one."

"Mr. Colonel, the same large river runs close to the village of Chornoliztsi, but these people were here, so how could they have beaten the Cossack? They are punished when they are innocent," said my father.

"Well, it seems that we made a mistake," he answered coolly.

"A mistake, indeed!" said my father, who sadly shook his head.

My father ordered a call to the rural police. He also ordered that the wounded be helped onto the village's carts and be taken to the field hospital in Tysmenytsya.

The colonel shamefully delegated two of his elders to oversee the task. The severely wounded were placed or helped into the carts, while those without severe wounds went on foot.

We would find out later that the Cossack was beaten by the boyfriend and two brothers of the raped woman in the village of Chornoliztsi and near the mill. On account of the Cossack's sly act, many people from three villages had suffered and in Chornoliztsi several farms had been burned down. Yet the colonel, it seemed, had admitted to a "mistake." From that time on, there was great hatred of the UNR troops led by Symon Petliura.

In a few days, my schoolmate Mykola Mikolyshyn, who served in the Petliuran Army as a division treasurer, came to us by car. He was in the rank of a centurion. I was very surprised that he went to the army because I knew that he did not like the Petliuran troops. I began to rebuke him for that.

"Do not be rebuke me, Mykhailo," he answered. "You have your own house and arable land, while I am a poor guardian's son. Where am I to go? What am I to do? My family is hungry, and so, in the army, I have something and the family is not hungry."

I told him about the recent massacre of our people and other crimes convicted by Petliuran soldiers.

"Eh, that's nothing. You should have seen what happened in Yamnitsa. You would not believe that that could have been done by *our own* people."

"What happened there?"

"There, the blood flooded the river! It was terrible! Would you like me to take you there? Only two hours."

"I don't know about riding. I have to ask my father. But tell me more. Your stories are more interesting than mine."

Mother gave us a snack, and Mikolyshyn began to talk. I lost myself in his story.

"In Yamnitsa, there were about two divisions of troops in combat readiness against the Bolsheviks, who had to cross the Dniester, but there were no exact data for that. The village of Sielets was evacuated the day before. Everyone went to Yamnitsa.

"At night, there was gunfire in a village between Sielets and Yamnitsa. The intelligence deployed there didn't find anything. The administration decided that this was a provocation by the Bolsheviks that was carried out by Yamnitsan peasants. You probably know that Yamnitsa was almost destroyed by previous battles, and the people who moved there from Sielets lived mainly in dugouts because they could not rebuild their houses and farms. But there was an order to punish the provocateurs, and punish them they did. They rushed like wolves—like *wolves*—at the defenseless population and beat up more than half the village. This 'mistake' was exposed later, but how—how can those crippled and mutilated return to life as normal? Who will heal the wounds of the wounded, set the broken bones, tend to the scars of the mutilated? If you wish, I can show you these beaten people."

"My father said I could go."

"You have a good heart! Well, if you want to go, I'll take you, but what will you take from that?"

"Maybe it will be useful for me," I replied.

"Well, then, we'll go, but we'll have to come back soon."

We wasted no time and left via cart immediately. The road was smooth and almost empty. In less than an hour we were in Yamnitsa, located in the valley above Solotvynska-Bystrica. Yamnitsa could not be called a village. It looked like a settlement from the Paleolithic age. Here and there, there could be found a hole in the ground, a dugout, and on the streets, puddles of blood were still visible. We saw two more unprotected corpses with heads busted open. Somehow, like a beggar, an older woman, having heard the rumble from the motor, left her dugout and looked at us.

"Do not be afraid. We are fellow Ukrainians," said Mikolyshyn.

"Hah! Those who ransacked our village said the same—that they are of our own, but you *see* what they did? Dear God! When will our grief be broken?" Her words were like frost penetrating our skin.

"Enough, Mykola! I have seen enough!" I said. Mykola told the driver to wrap up and return to Khom'yakivka.

What I saw in Yamnitsa was a nightmare that one could not forget. It was late in the afternoon when we returned to Khom'yakivka. I bid Mykola to stay the night with us and leave in the morning, but he politely declined because he said that he had some duty to perform that could not wait.

When Mykola left, I told my family of the destruction I had seen.

The day was exhausting, though when it came to an end I could not fall asleep in the barn. I rocked from side to side for a long time in the hay and did not fall asleep till sometime in the morning.

Thankfully, the Bolsheviks never came, but we suffered more than if they had come. The Ukrainian Petliuran soldiers behaved in our villages as if they were new occupants in a conquered land.

The front again moved to the east, and after the Petliurans, there came the Poles, who plundered and plundered again. They exhausted our collections of corn, hay, and straw.

In the fall, a policeman came to us and demanded four carts that should go to Tovmach. My father said that we could not give four carts but he would give two, with his own horses, because there were no horses in the village. I had to go with the policeman because my father did not trust my younger brother Volodymyr. And so I went with the policeman to Tysmenytsya, and from there we drove whatever carts we could gather, some ten, to Tovmach.

We arrived at the house of Dr. Ivan Makukha. From there, Polish soldiers took four sealed chests, which they placed on our cart. What was in those chests I did not know, but the people who gathered around the house said that there were in them seven or eight million Ukrainian hryvnias, which the Polish authorities had placed in the house of Dr. Makukha.

We were then ordered to go with two policemen to Stanislavivka, which was now under Polish yoke, and the Stanislavivkan police took those chests to the post-office.

We then were ordered to return to Tysmenytsya. Along the way, I wondered why such a sum of Ukrainian money should have been in the house of Dr. Makukha and whether every Ukrainian secretary-general had similar sums in his house.

I returned to my house late at night.

# Chapter 26[73]

## Recovery of the Reading Room

IN THE AUTUMN, I Began to take measures to restore the Prosvita Reading Room in the village. The reading room was destroyed in 1915, and its books were burned, and after having been somewhat restored, a family had come to live in it. With the family now gone, there was only one large oak table in the room. There were no lamps, no benches, and no armchairs. I had to come up with a plan to acquire all such things without money because there was no money for the project and no villager was willing to contribute any money. Polish paper money was in circulation, but none of the peasants were willing to accept it. The Ukrainian hryvnia was worthless, especially in the market. The hardest currency among the peasants was the Austrian crown, and it was impossible to buy something with it.

Having gathered together some fifty young volunteers, girls were instructed to whitewash the walls, both outside and inside, and to fill in holes in the walls with clay. I also asked each member to contribute, if possible, a bench or an armchair. My father also contributed some benches, which he made with his own hands from old boards. I bought bookcases and a lamp with my money. A stamp was purchased for membership tabs.

I had asked for advice from the main Prosvita Society in Lviv—a group dedicated to the promotion of reading. From there, I received an encouraging, even praiseworthy, reply. Yet one of the wicked villagers reported to the Polish police that there was some kind of suspicious activity in the reading room. Two policemen came to investigate the case. They said that I needed a permit, to be acquired from the Polish authorities, for restoration of the reading room, and without such permission, no one was able to proceed with the project.

My father and I mocked the policemen, and we told them that the activity of the reading room was based on the old Austrian law, which was still valid. Consequently, there was no reason to be suspicious or mistrustful, especially since the activities of the readers were to be socio-economic and educational, not political. "As a government official," said my father, "I am sure that no sabotage in the village has been conducted, so I beg you, be calm!"

The policemen, who had little understanding of the laws, were angry, but they retreated, and we proceeded with our work.

The Prosvita Reading Room was restored on November 7, 1920. Fifty-eight people became members, including teacher P. Petruniv, Father N. Ogonovsky, my father, and several other important patrons. However, few of the youth who signed up were active. In order to encourage their participation, I organized the Reading Amateur Theater Circle, which gave two small performances that winter, with the arrival of a few books purchased for the reading library. This event and other incentives did little to entice the youth to read. They preferred, especially on summer evenings, to gather in the square along the River Stremba and sing songs with great emotional significance.

I sometimes joined them, if only to get better acquainted with them and to try to lure them to the room. Such musical gatherings often lasted until midnight and were thereafter usually disbanded by Polish policemen. Of such unwelcome interventions, the elder peasants said, "The leaves on the willows were torn."

I suggested to those gathered at the River Stremba that they consider gathering in the village's reading room. Yet there was little enthusiasm for my suggestion. It came to nothing.

In the end, the Prosvita Reading Room, with its fifty-eight members, usually had five or six people for events, and never more than ten. There was no one qualified to give lectures on broad and relevant social and educational issues. It saddened me that all of my work had had little influence on the villagers, especially the youth.

It came to my attention that one young person who was walking home from the reading room late at night was beaten by Polish policemen. It often happened that policemen would harass and beat anyone on the street late at night. I told the policemen that no one who could prove he was leaving the reading room at night should be harassed or beaten.

I said that because of one particular event. One evening in May of 1921, four Polish police officers brought eighteen young men to our house. They were said to be disturbers of the peace. They were placed in a row in our house, and every young man was given seven or eight blows to his face. Blood flowed on the floor of our house and even sparkled on the walls. My father's intervention, at first, did not help. The policemen were angry because the young men resisted arrest, so they wanted to take revenge on them. The commander of that team, Shulsk, shouted that he would arrest them and lead them to the jail of the police office.

My father said to me, "It's a pity for us to witness innocent youth, though stubborn, being beaten." My father finally shouted, "Enough! I, as a government commissar, order you to stop this cruel behavior in my house!"

Then my father told all the boys to go home. He turned to the policemen and spoke in a manner that was sure to confuse or frighten them about threatening to report them to the county administrative authorities for abusive behavior. The policemen slunk to their office, and Father spat on the ground on which they stood after they left. He then sat down at the table and thought.

"See, Daddy?" I said with my breast swelled with pride. "What would have happened if you did not take the job and if a Pole sat now in your place? Those eighteen young boys would've been beaten half to death. Perhaps some would've been beaten so badly that they would've become crippled—crippled for their entire lives. No Pole would've come to the defense of those boys the way you courageously did!"

"Thank you, son," my father said in heartfelt fashion. Still, he was disturbed. He turned his back to me, sat down near the table, and sunk his head into his hands.

Three days later, three policemen came to our house. I brought up the reading room. I proposed to them that no one leaving the reading room late at night should be in the least harassed. They could do what they wished with others passing through the village late at night, but villagers leaving the reading room were not to be bothered.

The policemen agreed to my proposal. I then wrote up the agreement, and two copies were to be distributed to every Polish policeman who patrolled the village at night. Thus, the Poles allowed our villagers access to their native language in the Ukrainian reading room so that they could learn something instructive to pass along to others while they worked in the fields. After the agreement, the number of villagers who visited the reading room increased.

# Chapter 27

## The Cooperative

ON NOVEMBER, 16, 1920, I ORGANIZED the Constituent Assembly of the Cooperative. At these meetings, the Ukrainians from the Tysmenytsya Masters—J. Grinovsky and P. Rozdolsky—were invited to our village so that they, as "strangers," could explain to Khom'yakivkans the aim of the cooperatives. Father M. Ogonovsky, our new pastor, assisted me and cordially encouraged those in the church to come to the assembly and join the cooperative.

My brother Volodymyr went to Tysmenytsya to pick up the two masters. The two men, on the trip down, expressed some contempt for our villagers, whom they thought ignorant and indifferent to Ukrainian affairs.

When they arrived, my brother told me about their contempt.

I greeted the men and told them to keep an open mind. Not all our villagers were ignorant and indifferent. I asked them to remember that some people in the village were as educated as were they, so I asked them to treat Khom'yakivkans as equals. I said, "I have invited you to invigorate and raise up our villagers to a higher level of awareness. You, as intellectuals, can show them that all Ukrainians must work together to improve our economic condition. Our future lies in cooperation."

Grinovsky looked at me and said nothing. Rozdolsky spoke. "But Mr. Chemny, how do you know what our instructions have been and that we aren't here for some other reason?"

## The Oath

"I know what your instructions are, don't ask how. Please take every precaution so that your mission is successful. Otherwise, I will be forced to send you back without your speeches. Please, do not disappoint me!"

"Fine!" answered Rozdolsky.

The meeting went well. The gentlemen from Tysmenytsya behaved honorably.

The cooperative consisted of thirty members and had fifty-four thousand Polish marks. The directorate comprised Vasyl Petrishin (clerk), Semen Kalynchuk (secretary), and Mykhailo Chemny (clerk). Each of us donated ten thousand marks. The Supervisory Board comprised Father M. Ogonovsky (chairman), Kornylo Chemny (deputy chairman), and teacher Pavlo Petruniv (secretary). Other members were Vasyl Shkvarok, Semen Petryshyn, Vasily Yasninkovsky, Yurko Yatsura, Fedir Tretyak, and Lev Semanyuk.

The cooperative aimed chiefly first to serve its members with necessary goods that they could not get cheaply, but all the goods had to be acquired through the district cooperative—the Ukrainian People's House in Tysmenytsya, which was located in a public house in the middle of the town. I bought the boards for the house's cabinets, which I made with my own hands. The District Cooperative helped us with books. The cooperative, under the name Zluka, was registered in the district court after the claims of the authorities.

After two weeks at the Main Assembly, the Zluka cooperative opened its doors to the Khom'yakivkan peasants and even to those villagers from Odai, Slobidka, and Chornoliztsi. I also made a signboard. Having obtained from my father selected boards, I made an inscription with blue and yellow paints. The Polish authorities demanded that the inscription be in two languages, Ukrainian and Polish, but I made both inscriptions in Ukrainian. That got me into trouble with Polish policemen, who pulled the signboard from the wall of the house and said to us, "What does that mean?"

I replied harshly, "What right do you have to take down this inscription?"

"This is Polish territory! It should be white and red, not blue and yellow!"

"Why?" I asked. "After all, out of respect for your demands, I have included a Polish inscription."

"Who made this inscription? Hmm. How strange," he said. "The gentleman doesn't even know what inscription should be on the government store!"

"Sir, there is no government here! The inscription has nothing to do with the Polish government. It's for our peasants who know only their own language. I bid you to take this signboard and hang it where it was. Otherwise, I will be compelled to file a complaint against you for the disfranchised disturbance of someone else's property."

"Only if the paint is changed to white and red!"

"Well, I can do that, but give me paint, because I don't have such paint. Till then, I ask you to put the sign back where it belongs, and when I get the paint, then we can do what you want."

The policeman was angered. "Sir, I have the desire to arrest you for your insolence!"

"Is everyone crazy? Are you?" I replied with greater insolence. "And before you arrest me, please keep in mind that I am also a government official of some rank, and so it won't be so easy to arrest me. As an official,

# The Oath

I don't have to obey you unless you are acting under the order of the supreme authority. So, put the signboard back in its place and forget about the whole business."

The policemen did not know how to reply. They merely took the sign back to the cooperative and did not hang it where it used to be but placed it on the floor, where it leaned against the wall. It was placed again in its proper place by one of the cooperative's clerks.

The cooperative gradually grew and began to help many more Ukrainians. More members joined its ranks. The problem we were facing was depreciation of the Polish marks. Still, from the initial investment of fifty-four thousand Polish marks, the cooperative grew to over two million at the end of one year, though the two million marks had much diminished purchasing power.

Due to the high demand for goods, we had now fewer goods than we did at the beginning, so it was necessary to keep a close eye on matters each and every day because the prices of goods were in constant fluctuation. There was no place in our books for those uncertain numbers.

The depreciation of Polish money threatened the cooperative. The worth of our earnings from the sale of goods also depreciated. The Supervisory Board convened to discuss the problem but was not able to advise anything. It decided it was best to conduct business as usual and hope that Polish currency would strengthen.

Thus, the cooperative merely endured the dubious state of things till the end of the year. Still, the annual report showed that the cooperative had three million in assets, but there were not enough goods.

Clerk V. Petrishin went to live in America, and his brother Semen took his place. Semen was a stupid man, almost illiterate, and he, in his effort to "rescue" the cooperative, brought a few bottles of vodka to the store. I was furious, and I told him to return the vodka; otherwise, I would dump it on the street. And so a dispute arose between us. I threatened to make a complaint to the court, and that could lead to the closure of the cooperative, and he would be punished.

Semen took the vodka but decided to strike back at me. He hired several thugs to beat me at night when I returned home from the cooperative. I was informed about his role in the matter. Thereafter, I was careful and even got a revolver to protect my life.

On one cloudy night, when I returned to my house from the cooperative, where I had been recording and transferring accounts, four gangsters attacked me and knocked me to the ground. They circled over me—I saw merely blackened faces against the background of a gray-black sky—and I pulled out my revolver and fired shots at the feet of my assailants. Frightened, they fled. I fired again. After that, no one again bothered me.

When I was back in the reading room, I reproachfully told those present what had happened to me and that my actions on their behalf posed a threat to my life. I told them I planned to quit my work with the cooperative and the reading room. I worked for no personal benefit, but only from my love of and goodwill toward them.

The threat was availing. Some members told me they appreciated my efforts and that they would see to it that I was protected when traveling home. My public and cooperative activities demanded enormous effort from me. They also demanded that I travel to many villages: Tovmach, Tysmenytsya, Stanislavivka, Buchach, and even Lviv, far to the northwest. In addition, I had to go to such villages with reports, which made me vulnerable to Polish policemen. In order to distract the Poles and not to suffer from their chicanery, I thus hired a secretary in several villages, but that did not last long.

# Chapter 28

## Polish Appropriation of Our Lands

BECAUSE I WAS OUT of the house almost every day, my good mother, using her fine female instincts, advised me to get married. She even found a girl for me: Miss Theophilia Tymchyshyn. I liked Theophilia. She was not only hardworking but also beautiful. We were introduced and soon fell in love. On November 7, 1922, we got married.

Our young life was sweet and happy. Even before our wedding, my future father-in-law, who was in America, in Detroit, gave us his blessing and promised to assist us in any way he could. The promise filled us with hope for a brighter future somewhere abroad—perhaps in America—because life under Polish yoke was insufferable.[74]

The Poles appropriated our lands and began to farm them as if they were their own. Not waiting for the decision of the Council of the Ambassadors concerning Ukrainian territory, they were convinced that our land and its people were eternally their property, and so they treated us as their subjects and with great contempt.

# The Oath

**Picture 28.1. Wedding Picture, Nov. 7, 1922**
(Photo is painted.)

On December 21, 1922, Polish officials called forth all the troops of the last three years for active duty. I was among those called.

We arrived at Tovmach, where some two thousand of us, prodded by bayonetted rifles, were led like cattle to a nondescript schoolhouse. In the afternoon, I was inspected by the commission, and I was assigned to the 3rd Company of the 32nd Infantry Regiment. In the evening, we were led by bayonets to the railway station and driven into the freight wagons. The destination was Buchach, some eighty kilometers east-northeast of Khom'yakivka.

At Buchach, we were housed in the barns at the Ukrainian Gymnasium of Basilian Fathers. We lay spread out on the ground, like cattle on straw.

On the second day, we were led across the Strypa River, where Falcon Hall, near the Polish church, was located. There, we were kept for two days and then led by bayonets onto the freight wagons of the train station. The doors of the wagons were locked from outside, and at each station, we stopped at our doors, where a military police officer with a rifle stood on a rack. The Poles did not trust us and were afraid of rebellion or mass escape. We did not know where we were going because we could not see anything.

We were treated like animals. It was winter, and although there was a small iron oven in each wagon, it was still frigid outside, and there were forty people in the wagon, each vying to get close to the oven. For nourishment, they gave us a little potato soup, which nobody in our wagon liked.

In three days, we arrived at Modlin, a former Novo Georgy fortress that was some thirty-five kilometers northwest of Warsaw. Here, we stayed in the underground like ants for a week. Once a day, we went out to the courtyard, and there, behind a fence of iron bars, we saw many prisoners of war. Between them, there was housed the rector of the spiritual seminary, who would later become the Archbishop and Metropolitan of Ukraine, Constantine Bohachevsky. We could not exchange more than a few words with the prisoners because conversation was prohibited. It was a situation we had to endure.

From Modlin, most of us were taken to Dzialdowo and placed in former German barracks. Here, we began training and, against our wills, fell into a new, hateful life.

In two weeks of our life in Dzialdowo, we got our "military" clothes and a meager salary. Each of us had to sign for what he had received. I somehow carelessly put my signature on the wrong document, and Sergeant Dlugosch immediately drew my attention to that.

"What is your education?" he stated.

"Several classes of the Gymnasium," I answered. "You have my document. You must know."

"Please, go to the office!"

I went to the second room, where Lieutenant Nyzylkowski made me write a submission to the military district. I wrote what he ordered me to write.

"Fine!" he said, looking at the paper. "But you still have to go to training."

Sergeant Dlugosch—a malevolent, haughty, vindictive, and unintelligent person—called me the next day to his room and ordered me to clean his boots. I told him that I had never had to clean boots, but I would try.

With an arrogant smile, he admired my lowly work. "Well, well, what does this look like—a Ukrainian officer brushing the boots of a Polish officer?"

"Well, sir, life brings with it unexpected surprises, even reversals. Now I am cleaning your boots, but tomorrow you might be cleaning those of someone else. None of us knows what life will bring."

He gave me an angry look but said nothing. When I finished, I asked if I could go away.

"Yes," he said with resignation, not with contempt, as if I had given him something to ponder. He never again called me to polish his boots.

We were not used to the raw, bitterly cold air. Many of us got sick. I was constantly cold, though I never became seriously ill. Yet I did manage to have large furuncles over parts of my body, and that immobilized me. The civilian doctor who served our battalion sent me to the headquarters of the regiment, located in the town of Pomyahovka. During that time, I received a notice which said I was released from military service for two years. At Pomyahovka, I received no medical attention. I did not even get ointment for the furuncles that began to spread throughout the body.

On March 15, 1923, I was sent again to Dzialdowo, and there I learned from my comrades, among whom I already had by then many good friends, that I had a certificate ordering me to return home, which was now at the cooperative in Tysmenytsya. During my absence, my suitcase was stolen and in it, my smart uniform, with black clothes and boots. Only my cloak was left to me. I reported that with my orders to return home, but Lieutenant Nyzylkowski said that there was nothing he could do. Sergeant Dlugosch agreed.

"Sir, I do not have clothes to wear on the trip home."

"What am *I* to do? I have no uniforms to spare and no civilian clothes for you."

The instructor then gave me a piece of paper, on which it was written that I had received a leave of absence from the army for two years, and a card that allowed for all expenses paid in my removal to my house.

So I left the office and went to my barracks. A fellow soldier and friend from school gave me his civilian pants, which—though worn—still managed to cover my legs. Stepan Gryb from Koropets gave me his boots, but they were both too big and, consequently, very uncomfortable. To him, I gave my suitcase.

So I dressed myself, though the clothes I managed to scrounge up barely covered me, and went to those soldiers who were slated to take me to the railway station for my trip to Modlin. As I was covered in tatters, the cold wind drove through me. Still, I happily endured the bitter cold with the knowledge that I was going home.

The train reached Modlin in the evening. My case was settled by two clerks, one of whom, without rancor, iterated that I was going home. He seemed a poor fit for his underground office. It was too late to go to the station to wait for the morning train, so I, hungry and tired, found an underground cell in which to sleep.

In the morning, I traveled to Warsaw. The trip was slow. At each station, there was an inordinately long delay.

I arrived in Warsaw the next day. When I learned that the train to Lviv had already left that evening, I went into the city with the hope of finding something to eat from a generous soul. In the city, I entered several restaurants and said that I would work for food and some pocket change, as I had no money. No one helped me, and one gentleman even said, "Mister, I work for a few pieces of bread."

My Polish was good, so no one suspected that I was Ukrainian. Still, I got no food.

So I went to the Old Town in Warsaw. The Old Town comprised mostly Jews. The Warsaw Jews were different both in dress and mannerisms from the other Jews who lived in Poland. They wore long dressing gowns and yarmulkes with a small dash on their heads so they were easily recognized from afar. I struck up a conversation with one middle-aged Jew in the street and asked him where I could get a little food, since I had not eaten for three days.

The Jew closely examined me and then asked me to go with him. He brought me to a very dilapidated store and said something to the owner, who ran to me and said, "Oh, God! Mister is hungry! I am at your service." In a minute, he brought me bread and a few pickled herring.

The Jew who brought me asked if I wanted a smoke. I said, "Yes, but I haven't any tobacco." He ran into the street and in a few minutes brought me a quarter pound of good tobacco, paper, and matches. I ravenously dug into the herring and bread, and the storeowner brought me warm coffee in a mug. I cordially squeezed his noble hand and thanked him with a sincere heart.

They asked me where I had come from and where I aimed to go.

I did not dissemble to these good-hearted people. I even admitted that I was a Ukrainian.

"Oh, good God! You are still far from home!" The owner wrapped some bread in a piece of paper and gave me the small sum of one hundred marks for the road.

I thanked both cordially because I found in them the Good Samaritans from the Gospels.

Leaving the store, I once again thanked them and said goodbye.

They answered me, "Let the Lord guide you on your return to Ukraine!"

These good people were not Christians, but they helped me in my time of large need. Thus, these poor Jews, through their kindnesses and benevolence, were a hundred times better than their Polish neighbors who called themselves Catholics.

The train I rode from Warsaw to Lviv was much faster. In the morning, I was already in Lviv.

I wanted to see the city, but the train to Stanislavivka would be leaving in a half an hour, so there was no time for sightseeing. That proved beneficial because, when I examined myself in a mirror at the station, I saw clearly that I looked the part of a vagrant—not a soldier or son of a commissioner. I contented myself with a cup of coffee for fifty marks and ate the bread that my Jewish friends in Warsaw had given me. That improved my mood.

Late in the afternoon of March 21, 1923, I arrived at Tysmenytsya. I jumped from the train and ran on the winding road that led to my father's residence as deputy chairman of the cooperative.

It was twilight as I entered my father's yard. My father was in the hallway, and when he saw me, he cried for a while. We cordially greeted, but when my father saw my sorrowful look, he cried even more. Then my mother, my wife, and my siblings greeted me.

I was yawning and sleepy, and I could not much manage my left arm because of two large boils: one at the elbow and the other on a shoulder.

"But weren't you fed by the Poles?" asked my mother, who rushed to warm a cup of milk for me.

My father saw my boils and said, "Tomorrow we go to the doctor! We have to do something!"

The next day, we were in Stanislavivka and visited Dr. Yanovich. The doctor examined me and said that he found the boils overwhelming. He then prescribed medicines, ointments, and bandages. We said goodbye to him and left.

For nearly half of the summer, I was treated for my boils, which had spread to most of my body. Yet the medicine much helped, as they were no longer so painful.

At the same time, I was engaged in paperwork in the community and reading room. Our cooperative, during my absence, had degenerated. The devaluation of Polish currency took a toll, but so did many other blunders.

The visiting inspector from the Ukrainian Revision Union from Lviv threatened to close the cooperative. He called an emergency meeting in the chapel near my father's house and affirmed negligence on the part of the directorate of the cooperative. I took no active role—first, because I was ignorant of what had been going on for several months, and secondly, because I needed rest from the many months in the Polish Army. Yet the auditor voiced some hope for the cooperative.

"Things were going swimmingly when Mykhailo Chemny was overseeing events. If this gentleman," he added, "will not assume the same role as he had before, then I have no choice but to close the cooperative. It has no right to exist."

My father, who was at the meeting, was pleased and proud to hear from the mouth of such an esteemed man such praise for his son.

The auditor as well as the chairman of the Supervisory Board, Father Ogonovsky, visited me at my house and beseeched me to assume my former role in the cooperative. I listened to them, but the work was hard, and there was no promise of success. My co-workers were unwilling to listen to me, and there were few buyers of goods.

Yet I agreed to resume control of the cooperative. With hard work and persistency, I succeeded in pulling it from the foul waters in which it had been in my absence.

Still, I was unhappy about the state of affairs in my beloved Ukraine. So, when I had free time, I sought out ways of escaping Polish yoke. Through immigrating to a non-oppressive state, I could finish my studies and cultivate a freer life.

For that purpose, I submitted an application to the regional headman for a passport for departure abroad. It was categorically denied. The commandant of the police station in Tysmenytsya, who investigated my submission, even threatened that if I would try to leave, he personally would oversee my case to keep me from leaving. He stated, "Let the gentleman use his head to free himself from that state of affairs!"

Before the Christmas holidays, in early 1924, my mother-in-law, Mrs. Tymchyshyn, entrusted me to change twenty dollars, which her husband, Joseph, sent her from America. I exchanged the money on the black market.

There, I met a Jew named Kimmel. In his home, he crafted a plan for me to leave Polish Ukraine. He promised to help me, to get me a passport and other needed papers, but the passport would cost fifty American dollars. I agreed to all that, and I acquired a lasting contact with Mr. Kimmel.

My wife's father sent one hundred fifty dollars for me to give to my mother-in-law.

On January 28, 1924, Theophilia gave birth to a daughter, whom we called Eustaphie. It was a happy and gratifying moment, and my father-in-law, proud of his first granddaughter, asked that the money be given to me. In order not to take money from my mother-in-law for nothing, I promised to build her a large barn, for which she had for years collected materials. I completed the barn early in April.

# Chapter 29

## "Count Osaslavsky ... wants to see you"

ON FEBRUARY 15, 1924, There was a terrible and thick snow. The members of my family were sitting in our warm house, and each person was busy with some task. Leaving the house was unthinkable because the weather was worse than bad.

Near noon, there was a knock on our door. A messenger covered in snow greeted us in the Christian manner and said, "Count Ososlavsky, who is now in Pshenychnyky, wants to see you!"

"In such a blizzard?" I asked.

"Yes, the weather is bad, but the count wishes to see you and has sent horses for you."

I could not guess what Count Ososlavsky wanted from me, but I agreed to go and threw on my fur. Pshenychnyky was not far away, only three and a half kilometers, but a trip through such a blizzard, though short, would not be pleasant.

On the way, I began to think about what the count might want from me. I had not seen him since he graduated from folk school, where he, a young sir in the society of Prince Lyubomyrsky and the princess and their daughters, took his exams. In the end, I concluded that this visit to him was probably connected with some kind of tax. The Polish authorities had issued a law that each farm belonged to the Polish government and, at the same time, fell under the jurisdiction of the village in which it was. A few weeks before, we received a letter from the Treasury office asking about the farm conditions of the village Gorodishche. We did not know that the Gorodishchean officials had declared that their farms were destroyed by seventy percent from its pre-war status. They wished to avoid any taxes. In contradiction to their statement, I submitted my report that the farms were destroyed by thirty percent on account of the war.

With that in mind, I entered the director's office in Pshenychnyky. I was led into a beautifully furnished room, where at the desk there sat Count Ososlavsky, the son-in-law of Prince Lyubomyrsky. He stood up from his chair and greeted me. Yet he soon began to interrogate me, which put me, given the bad weather, in a mood worse than I was. To make matters worse, I had to turn back to Hierevorsk to attend to other business.

"How can I serve you, Mr. Count?" I said.

He bid me to sit down and asked whether I froze during the ride. He was readied to treat me to good cognac, but I refused. He then turned to business.

"Mister is a secretary in Khom'yakivka?"

"Yes."

"Why did the gentleman in his reciprocation file give a false report about Gorodishche farm?"

"Mr. Count, I gave only a file that condemned the headman and three members of the public council. I only occupy the position of secretary and nothing else."

"But your headman and his advice team understand nothing in this business!"

"Sir, I am far from being a judge! They made such a fuss, and so I lodged it with the Treasury and they signed it. Why didn't Mr. Count call on them?"

"Because they are stupid and do not understand anything, and I—I do not want to talk to them!"

"Well, then, what do you want from me?"

"I prefer to talk with an intelligent person, not with a rude man!"

"And when I play my own hand and not yours, maybe I can live up to my 'honorable' name?" I said, with an anger that began to swell in me.

"I ask you to be calm. I know that you are a nobleman, so I want to talk with you and not with anyone else about this matter. Please!" He paused and then continued. "You yourself know that the Gorodishche farms were destroyed more than thirty percent. The chamber is burned, there are no livestock, the field tools are not useful, and the stables are in collapse. Thus, the condition of their farms is destroyed by more than seventy percent, and you submitted thirty percent. By what logic, gentleman, were you guided?"

"Well, let's see. Once, the yard had fourteen pairs of horses, now it has nine. The cattle were thirty, but now twenty. It's true that the chamber is burned, but it does not belong to profitable items. The stables, too, are in a bad condition, but that is not the result of war but of neglect. During the war, no one took the initiative to repair them."

"You're smart!" said the count in Polish with some venom in his words. "But if the gentleman goes to our side and states that the farm has been destroyed by seventy percent," he began in Ukrainian, "then we will have you to thank, and there will be no harm done to you."

# The Oath

"No, Mr. Count, I cannot do that, because the request is already in the Treasury, and I cannot submit two different reports about the same thing."

"We can destroy the first report, and nobody will know about it."

"I do not know how the Treasury will look at such a manoeuver, but that is beside the point. As to no harm done to me—that no one will care if I destroy the first report and replace it with another—I dare say that there will be harm!"

"How is that?"

"Well, the Polish government and its Treasury have a budget, and that budget is already calculated and based on profits of the state. When some part of citizenry in some way escapes their taxes, the rest of the citizens must make up the difference. In other words, if you do not pay your measured tax, then I, and others like me, will have to pay it. Hence, changing the reports will bring harm to me."

He cunningly smiled and said, "Your calculating at least seems correct at first glance, but your reasoning is off because taxes have *already* been calculated for this year."

"I suppose so, but in the second year they can double."

"Dear sir," he began again, "I shall give you ten wagons of wood and pasture for ten livestock. If you do not need that much land, you can sell it and keep the money."

"No, Mr. Count, I couldn't do that, even if you'd offered me this farm," I replied coolly.

"Hmm, a proud Ukrainian nobleman," he said through teeth nearly clenched.

"Thank you," I answered, getting up from the chair. "I think that my visit is finished. Please kindly send me home."

The count's face at first turned blue and then got red with anger. After a few moments, he stated in Polish, "As you wish. I cannot do anything more!" He got up from the table, went to the next room, and told his director that I needed to be sent home.

I bowed down to the count and went to the cart to be driven home.

In early March of that year, the count's carriage came again to the front of our gate. The driver informed me that Mr. Director of the farms in Pshenychnyky wanted to see me on very important business.

"Which one?" I said.

"I don't know. I'm here to drive you."

I went.

Upon my arrival, the director, a Pole, greeted me on the doorstep of his apartment and took me to his living room. In the living room, he asked me to take off my coat and sit down. He apologized for not knowing sufficiently well the Ukrainian language and said that he would speak Polish, and I could speak to him either in Polish or in Ukrainian because he, though not fluent in Ukrainian, understood it.

"Well," I said, "let's respect both languages. Let the gentleman speak his tongue, and I, my own," I said, sitting down.

"Sir, forgive me for disturbing you, but Mr. Count asked me to bring you here, and so I did it. The village, specifically Ostrivina's farm, has no manager. Between the village and the farm, there are—well—the relations are bad. Recently, a barn was set afire, and last summer a lot of rye was burnt. And so—and so, would the gentleman not agree to take the farm yard into his own hands and establish relations between the village and the yard?"

"Mr. Director, I'm not an agronomist, and I don't have a farm board, and I have neither sufficient science nor sufficient experience."

"You are a son of the earth, and you know how to work the land. We know that you've studied economics for a few months, and that will get you by. We shall also help, as you will need our help. It is important that there be good relations between the yard and the village, and you, a Ukrainian, can do just that. Mr. Count promised to give you ten morgues of land,[75] building materials, land for grazing your livestock on the farmstead, and training for your children in the high school. Moreover, we'll pay you a good wage for your efforts. What do you say to that?"

"Mr. Director, all that sounds very good, but in practical terms, nothing can be done. The village has its own requirements, and in the yard, I will find myself between a hammer and an anvil. Serving two masters is probably not possible. I know that sooner or later the yard would require me even to do the Latin rites, and I could never do that!"

"And what if it came to that? What damage would that do?"

"Much! I will never be a traitor to my faith and my people for the sake of the good fortune of the wicked and for my own benefit. Tell me, by the mercy of God, what would you do if you were in my place?"

"That—that would depend."

"You see! You see, you are hesitant to give a clear answer," said I, "so *I* shall give you a clear answer. Under no condition and for no reward shall I *sell* myself!"

"But please, sir, this is a golden opportunity—one which happens once in a lifetime and not for everyone. The gentleman is a lucky man! The circumstances of our lives are not so rosy. The gentleman will not get a better offer!"

"Thank you very much for the proposal, but I don't want to be a servant all of my life. I want to live honestly, even if that means living poorly!"

"Where does the gentleman find such a life in the present times?"

"I'll go to emigration!"

"But there, too, gold pears do not grow on the willow."

"I do not expect gold pears, but I want freedom! I've never been a slave, and I don't now want to be one!"

"Does this mean that there will be 'no bread from our flour'?"

"No, it doesn't!" With these words, I got up from the chair and went to take my cloak.

"Sir wants to leave?"

"Yes!"

"I'm sorry that the gentleman did not accept our offer. I suppose that the gentleman will regret it—regret it someday."

"Maybe, but I doubt it. Can I ask for horses to go home?"

"But I offer you very much!" He paused. "Hmm, the horses are ready."

He saw me to the door and nodded for the rider to take me home.

# The Oath

When I came to my house, I spoke about this comedy to my father and the rest of my family. My father pondered a bit over my story and said, "Though it was a seductive thing, I'm proud of you—proud that you did what you did. I don't want you to be a Polish slave for your whole life, even though others would call you a lord."

# PART IV
# HAVANA, CUBA

# Chapter 30

## "It's hard and painful for me to part with you"

IN THE BEGINNING OF 1924, My father, in bad health, abandoned his commissariat, which he always detested, and gave it to one of the peasants, a certain Semen Petryshyn, who very much wanted the job. It was a good fit for Petryshyn and the Poles. Petryshyn much respected the Poles, and they liked him more than they liked my father. Yet Petryshyn was called sometimes to his face *hrun*, a fool, as he was a servant of the Poles. He did nothing for the people but merely avenged himself against those who offended him. And so, once again, the policemen were daily guests in Khom'yakivka, where my father had returned, and they beat up and bloodied the faces of innocent people.

Conditions of life in the village became intolerable. People came to my father and complained. "Why did you abandon your position and put us in the hands of such a fool?"

Many from Khom'yakivka and neighboring villages went to France to escape from the hopeless life in Ukraine. There were days when I was producing five or six applications for passports abroad per day. Everyone wanted to leave, so great was the abuse and inhumanity of Polish officials. The appointment of Petryshyn was not without implications for me. There was concern that Petryshyn would replace me with a Pole from the city whom the peasants would be forced to bear.

# The Oath

I, too, recognized that I had to leave. My father and I tried to get some money for a trip abroad. One Sunday, we went to the neighboring village of Markivtsi to see Nastyuk, who had returned from America and loaned us three hundred dollars. Mykyta Semanko, who had also returned from America, loaned us two hundred dollars. The rest of the money came from my father's savings. In all, we collected around one thousand dollars. As collateral for the borrowed money, my father guaranteed all his property.

What would happen if my trip failed? I tried not to think of that, as I would be failing my father, and it was not easy to round up such a sum of money. Yet my father had such a deep trust in me that he did not hesitate to do anything but to help me. He said, "Son, it's hard and painful for me to part with you, but I sincerely wish you well, and I'll do whatever I can to help you. I hope you understand that and that you won't let me down."

"Never, Daddy," I said. "As soon as I can, I shall pay back everything. I trust the Lord God, and I believe that he will help me!"

I told my neighbor Demyan Kalynchuk, who had a wealthy brother in America, about my plans with the hope that he might be able to help me if I needed his help. That was a mistake. The boy, unkempt and arrogant, promised no monetary help, though I, with the assistance of the Ukrainian Commissar, Mr. Chaplinsky, did much to help him. That said, Demyan would be my traveling companion on my trip.

Although the villagers were not at all grateful for the work I had done, I loved Khom'yakivka for some reason and I would be sorry to leave it. Yet I also realized that I could no longer be of much help and that staying longer was not an option, so great was the degradation.

I asked my Jewish sponsor, Kimmel, to accelerate my departure. He promised to do everything possible. Within a week, I got a letter from him. He said that he wanted to see me.

I came to see him the day after I had gotten the letter. He said that everything was ready and told me to be at the railway station in Stanislavivka on Saturday, April 16, at five o'clock. On Thursday, I told my headman that I would be leaving for Warsaw for several days to fix some affairs concerning the departure of my mother-in-law to America and that if there were any pressing cases, the teacher Petruniv could handle them. My mother-in-law's departure was already common news in the village, so it was easy for me to cover my departure.

April 16 was a very sad day for my family and me. With tears in her eyes, my wife, Theophilia, bid me farewell. My parents' sorrow could not be abated, as they in their heart felt that they would never again see me.

Taking with me only a small suitcase, and heavy in heart, I traveled toward Tysmenytsya. I was giving up my life for a future that was unknown. I was anything by optimistic.

# Chapter 31

## To Paris, "Capital of the World"[76]

**Picture 31.1. Warsaw Railway Station**

MY TRAVELING COMPANION, DEMYAN, did not leave with me. That would have caused suspicion. He caught up with me far beyond the village. As he was a member of the headman's Brotherhood, it was even more dangerous to go out together.

At Tysmenytsya on April 16, 1924, we boarded a train and arrived at Stanislavivka at four p.m. Kimmel was waiting for us at the station. He bought himself and us tickets to Warsaw, and at five o'clock we left Stanislavivka.

On Sunday, late afternoon, we were already in Warsaw. Kimmel brought us to St. Yuri Street and also placed us in a friend's house—his name was

# The Oath

Friedberg—where he, too, stayed. Friedberg's house, though clean, had an abundance of fleas. At night, I could scarcely squeeze tight my eyes, because I was terribly bitten by the fleas and my face swelled.

On the second day, Kimmel wanted us to go to a photographer who was going to make us some passport photos. Yet it was impossible for me to get a passport photo with such a swollen face, so we had to wait until the evening when the swelling abated. We went to the photographer in the evening, when the swelling had lessened significantly.

There was an order in Warsaw that every new arrival in the city was to check in with the police. I did not know of the order, but I merely sat in the house, and I did not roam the streets of the Polish capital of such great bounty. It was good that I did not go out, because, as I later learned, our "good" headman had on Tuesday announced to the police that I had escaped and taken with me one of the Brotherhood, who was not yet of military age. As Demyan was considerably younger than me, I was wholly to blame for both of us, and chase letters were laid for me.

On Saturday evening, we received passports with a visa to Gdańsk—I, as a student of Polytechnic at the University of Gdańsk, and my travel companion as a worker. Both had Berest, near the West Bug River, as "Origin." My name was Maxym Mlinarsky. Demyan's name was Karol Kovalsky.

In Gdańsk, we had to get new passports to France, and in France again new passports to cover our tracks. In Gdańsk, we also had to get information about where to go and what ship we needed to board. I had addresses of various agents who apparently would help us on our journey in different places, such as Gdańsk, Le Havre, Paris, and San-Nazaire. It turned out that Kimmel had wide connections, and so I began to have great trust in him. On Saturday at eleven p.m., we took a *fiacr* (cab) to the Gdańsk railway station. Kimmel and young Friedberg went with us. They had planned a quick train in the first-class wagon for us, and they advised us on how we should behave so as not to arouse suspicion.

"Mostly act like you're sleepy," they said.

The first-class car was divided into two parts. In the first part, there traveled Marshal Joseph Pilsudski, who was going to his summer camp in Weighorod; and in the second, there were we and several other passengers. Kimmel and Friedberg rushed us to our car, because there were few police and no police checkpoint. Somewhere near Tschew, two policemen came to us and asked about our passports. Acting sleepy, we gave them to the policemen, who examined them, calmly returned them, and then went on their way. That brief inspection plentifully frightened me, as I fully expected that we would be found out to be frauds. I gave thanks to the Lord God that all this was so easily over. I came to realize that Kimmel and his accomplices were men of great courage and large souls.

When the train was on the land of Gdańsk, a free city, two German officials came to us. They looked at our passports and one of them asked in German, "Where do you think to settle in Gdańsk?"

"Mersier Strasse eleven," I replied as coolly as I could.

"Is there a dwelling place?" he asked.

"Yes, above."

They returned our passports.

It was already morning when we arrived in Gdańsk. Though I felt safer there, I still recognized that we could be detained, and after questioning, could be handed over to the hands of any of the Polish police who were numerous in Gdańsk. The worst thing was that my passport listed me as four years older than I was. Demyan was listed as nine years older. Each of us looked very young, and experienced eyes could see that.

After the train ride, we took a *fiacr* to a specified address. There, we were approached by a young man of Semitic origin and asked to enter. He told us that at noon a ship would sail to Le Havre, France, and if we wanted, he would get us tickets for the trip. At that point, I did not much care where I might go—I wished only to get farther from Poland—so I agreed. Demyan stared blankly at us because the conversation was conducted in German. When I told him about our trip to Le Havre, he said that he desperately wanted to see the city.

"You will see not one city from now on, but many. Now go, while there is a chance!" I said.

At noon, we paid for our tickets—five dollars for the service plus twenty dollars for transport—and were on the ship. The ship was full of many kinds of French workers. We traveled for several days across the Baltic Sea and then through the Calais Canal to the North Sea, until we arrived in Le Havre in France.

From Le Havre, we went by railway to Paris. The ride was long but nice. There were beautiful landscapes and pretty French houses, and such scenes elevated our mood. Everything thus far had been going swimmingly. At every step, however, I could not help but thinking, *What's next?*

The cramped and short French railroad wagons rolled and shook on the railroad track. Often one could openly see military carts, guns, and other military equipment. Those were not hidden, as they would have been in Poland. They were paraded as if to be witnesses of the recent war.

Late at night, we arrived in Paris. We settled at the station until the morning, when, after breakfast, we went by taxi to a nearby hotel. From there, we saw what we could of the "Capital of the World."

A taxi driver took us to an address I had been given. The man who met us—he called himself Stern—was obviously a man of "our faith." He told us that we could not expect to go to the United States or to Canada, because there were no tickets. Yet we could travel freely to Mexico, to Cuba, or to parts of South America. Having friends and relatives in the United States, I chose Mexico or Cuba.

In a couple of days, the sailboat *Spain* would be leaving San-Nazaire, some thirty kilometers west of Nantes. For twenty dollars, we got new Polish passports with new names and visas to Mexico and Cuba, where age did not make much of a difference on those passports. My new name was Aleksander Kolosovsky, and my title was teacher of Slavic languages. My companion kept his same name because he said that he had already gotten used to it.

I discovered that there was in Paris a club of former elders of the Moscow army who fled before the Bolsheviks. It was not far, and I decided to go there and learn what I could learn.

The club was located in a large stone hotel. There were many different types of Muscovites, most of whom wore their uniforms. Few were in civilian clothes. I sat down at an unoccupied table and tried to listen in on the various conversations, but I could only hear what was spoken at the nearest table, where four people were seated: a colonel, two centurions, and a lieutenant. Each was dressed in a military uniform with honors, as in royal times.

The waiter approached me and asked what I would order. I asked for a glass of wine. The wine tasted ill to me, so unaccustomed was I to drinking it. A fellow in civilian clothes near me said, "Good day," and then began asking me when I arrived in Paris, how long would I be there, and in what part of the military I had served.

## The Oath

Fearing exposure, I gave him fictional answers, which he accepted as truth.

He said that he served as a door supervisor in the hotel and earned little money, but he got along.

"I wish to leave France and look for better luck somewhere far away."

"Well, then, only in South America is that possible. There's nowhere else."

"But why?"

"Say nothing of this to anyone. But—but only from South America it is possible to get into the United States or Canada."

"How is that?"

"For some money, everything can be arranged!"

"Well, I haven't much."

"It takes only a little. I have already sent many to North America."

I finished my wine, bid farewell to my companion, and left.

I went back to the house where we were given passports, and there was Demyan. I asked Mr. Stern to buy us a map of Mexico and Cuba.

He then said, "It should cost two hundred ten dollars to travel."

"For a ticket?"

"No, for two."

"Well, that doesn't seem so bad!"

"Tomorrow, we will buy a map."

"Good. Thank you."

The next morning, Stern called a taxi, and we went to the ship's office and bought maps of Mexico and Cuba. The agent of the bureau, having inspected our passports, said that everything was fine and that we could go to San-Nazaire tomorrow because it was almost a one-day ride. We asked Stern to help us get on the train and to get us nonstop tickets San-Nazaire.

At seven a.m., Sterne came for us. He hailed a taxi, which took us to the railway station. There, Stern bought us two tickets, and we gave him five dollars for his service.

The train to the pier was quick. This track was much smoother than the one that led from Le Havre to Paris, and our train traveled at eighty kilometers per hour. The ride overall lasted some twenty hours.

# Chapter 32

## An Improvisational Ukrainian Concert

WE WERE IN SAN-NAZAIRE at about Eight A.M. on the next day. The railway station where we arrived was near the pier so that we did not have far to walk to get to our ship.

We went to the bureau of the shipping company, and the boss showed us to the ship. We left our heavy clothes in his office, because it was already fairly warm outside, and went out to see more of the city. We did our share of walking to view the magnificent city. When tired, we ate and then sat in the park near the pier. Thereafter, we returned to the ship, which would not sail till just past midnight.

At the dock, we went on the deck. There were two distinct views. On the one side, there was a beautiful yet calm sea. On the other side, there was San-Nazaire—a handsome city that was adorned in the fresh verdancy of spring.

We were called for dinner at around five p.m. I was transported by a fine mist on the deck that kept me from thoughts of eating and made me think of the possibility of a happier, free future. Still, I descended to the dining room if only from curiosity. There I had coffee, while Demyan had some soup with meat. Demyan and I talked much. At some point, a man in a blue American shirt came over and introduced himself as Petro Yatsina from Rava-Rus'ka, a village some thirty kilometers northwest of Lviv.

# The Oath

He said he was returning as an immigrant to America. He was accompanied by his neighbor, a female. I immediately felt comforted. Having a fellow Ukrainian on board would make the trip much more enjoyable.

After dinner, Petro, his neighbor Maryna, Demyan, and I went on the deck and found a spot where we could be by ourselves. There, we resumed our conversation.

It was a pacific spring evening, and my mood was fine. I, with a fine tenor, began a song, sung mostly for myself. Soon Yatsina and Maryna joined in the song, and even on the foreign, calm sea that nestled the setting sun, the soft Ukrainian song could be heard.

Hearing our Ukrainian song, Dmytro and Vasyl from Volhynie, just north of Lviv, joined us. And then, two from Peremyshlyany, some thirty kilometers east-southeast of Lviv, joined us. And more and more came, like hungry birds when somebody spills wholesome grain. Even a Pole from Zamarstyniv, on the outskirts of Lviv, joined us. He loved much the Ukrainian language but spoke Ukrainian-Polish. In all, we were a group of some fourteen people. The song was moving and evoked in each of us, I suspect, some pleasant dreams.

We were a happy but discordant band of singers.

"You must conduct us," Yatsina said to me.

"It will get better," said I with a hopeful smile.

"Yes, indeed! Watch out!" said another with a laugh.

And so, not knowing how or why, I became the first and only "conductor" of the improvised Ukrainian choir. I knew many songs—historical songs, songs of the homeland, and love songs. I gave the word and began the melody, and all of us were singing loudly: "Where the Dnieper is, we make waves...." The dulcet sounds—there were a few dissonances—not only reverberated to the shore but also echoed far along the smooth pleasures of the sea.

Many were now gathered on the beach and in the small park—the workday was over, and it was time to relax—and our song about the Dnieper warmed their hearts. When we finished, there was loud applause from the park, and that encouraged us to continue singing. I began the old Cossack song "Do You Grow Up a Tall Oak." That, too, received loud applause, and so we sang "Kozachenky," which, too, received loud applause. There were shouts: "Repeat! Repeat!" We repeated. Even the ship's captain left his work and listened to our singing.

As I write, I realize now how very foolish it was for me, a fugitive, not only to sing Ukrainian songs but also to conduct the choir. Yet my young, proud soul was inattentive to carelessness. Moreover, it was a joyful occasion, and I had had none in the last few weeks, and so I let my heart, too accustomed to suppression and pain, fully express itself. For once, luck was on my side, as the letters sent by the Polish police had not yet reached San-Nazaire.

When we stopped singing a little bit, there was unrest in the park and on the beach. There were shouts for us to continue singing. Even the captain, appealing to us in German, said we should sing something else.

"Our throats are too parched to continue, Mr. Captain," I said.

The captain whispered something to one of his officers. That officer went to the ground floor of the ship. In a matter of minutes, two sailors came in regal white uniforms with a tray, glasses, and a few bottles of good Portuguese wines. They filled the glasses and offered one to each of us. We gratefully accepted the gift and continued our choir. We sang "Chumak" and "Peasants Cry," and still other well-known songs. Most of all, they liked the song "Shirokeye Bolognachko Woda Zalyala," for which we received boisterous applause from what were now a few thousand people who gathered in the park and on the beach.

Judging from the captain's reactions to our singing, San-Nazaire was probably the first—and maybe last time—he ever listened and would ever listen to a Ukrainian song.

It was nine p.m. when we finished our improvisational concert. Some people wanted us to continue singing. Others, we recognized, were tired and needed to go home to rest. And so, enough was enough. Moreover, the captain had much to do. The ship was scheduled to depart precisely at five minutes past midnight. The captain went to his cabin in preparation for a long journey. Each of the members of our "choir" went his own way.

# Chapter 33

## Leaving San-Nazaire

WHEN THE CHOIR DISBANDED, I remained on the deck to see the ship leave port, which I was told was something I needed to see. I saw, as the hour approached midnight, sailors attend to various tasks such as securing cords, fixing benches, and rigging clogs of lifeboats.

At midnight, the ship's engine was thrown and the whole ship shook. From its chimneys, there rose a thick, black smoke. Then the anchors were lifted and the thick ropes that tied the ship to the pier were slacked. At the same time, the bridge was removed.

At five minutes past midnight, the ship pulled quietly and slowly from the shore. As it pulled from shore, it shined a huge spotlight on the ever-emerging sea. The light gradually was only visible as a sort of white bark under which was the black shore of the earth.[77]

Whoever has never seen a ship leave port at night, especially in the spring, cannot imagine the feelings of both beauty and sublimity that simultaneously well up in one's soul. The beauty I had felt all along, but as we parted, the sublimity overwhelmed me. The black of night, the endless sea would soon surround me from all sides. I looked for a long time as San-Nazaire disappeared, and soon, the coast of Europe disappeared. My soul, before elevated with song, was now laden with sorrow.

*My homeland,* I could not help but think. *Will I ever see you again?*

## The Oath

It was a long, eventful day. At about one a.m., I went to my cabin to sleep. I could not sleep. There were too many thoughts flooding my mind.

On the next day at around noon, we arrived at the Spanish port of Santander, where the ship docked until the next morning to load and unload cargo. The ancient harbor city was nestled around a beautiful mountainous bay. Because of the layover, whoever had permission to visit the city was able to do so. I took advantage of that opportunity and left the ship.

The streets were narrow and steep, with narrow stairs going to old stone houses that spoke of time long ago and of an old style of construction. On the streets near the pier, there were various small merchants, who peddled their goods with sonorous shouts. The sidewalks were filled with various kinds of big baskets of different vegetables and fruits. In the oldest streets, there were handsomely attired gendarmes. Overall, the city was orderly and clean and bustling.

Having a brief tour of the small city and having bought fresh cherries and other vegetables, I returned to the ship. I noticed then many merchants, especially peddlers of foodstuffs and some in small boats, who crowded around the ship. The purchased goods were wrapped up and pulled up to those on the ship with the help of a lace, while buyers tossed coins from the deck to the merchants below. The scene was somewhat comical from afar, as many of the tossed coins found their way not into the baskets or boat of the merchants, but into the sea.

In the morning, we headed west and at night sailed to the Spanish port of Coruña. We rested there until morning and then again took to the wide sea and headed south. To my left, I saw the coast of Portugal, and from the right, the wide, infinite sea.

In the evening, we arrived at the Portuguese port Porto. The big harbor city, spread over mountainous terrain, did not especially attract me, so I remained on the ship and studied it from the deck. A high mountain to the right of us, however, blocked us from much of the city, so there was a limit to what I could study.

We remained in Porto for a day. The Portuguese regaled us on the ship with their chanted pleas on behalf of their goods, yet there were few buyers among us. All of us had far to go, and most of us had little money, and so we had to spend our money judiciously, not liberally.

**Picture 33.1. Ship Leaving San-Nazaire (1940s)**

On the evening of the next day, we turned toward America and away from Europe. The ship turned southwest and began its trip across the wide Atlantic Ocean.

The shores of Europe, disappearing from my eyes, seemed to die. That, at least, was the impression I had, and I am certain that most others had,

# The Oath

as leaving the coast of Europe was a bittersweet moment—a moment of soulful sadness yet quiet optimism. Willfully leaving our homes with hope of a better future, we were leaving behind everything we had known and had ever had. It was the death of one person and the rebirth of another. Yet the uncertainty of the future was terrifying. With the turn of the ship to the southwest, there was no possibility of a change of mind.

# Chapter 34

## "I have a visa for Mexico and for Cuba"

THE FIRST HALF OF the day on the wide ocean was not bad, but the rest of the day was unpleasant. The waters, aggravated by a merciless westerly wind, were rough and so the sailing, too, was rough—rough and slow.

Demyan and I went to the deck, but the waves, crashing against the ship, shot briny water through the fence and seemed to aim to wash away everything that was not anchored to the deck, and so we returned to our cabin and tottered on our beds while we watched others passing the time. Some played cards or chess, while others were sitting in chairs or lying in beds and doubtless dreaming about what was ahead.

Headed for Cuba, I knew relatively nothing of the language, so I bought a German-Spanish dictionary and tried to learn Spanish. Having learned just a few words, I realized that this would not go far, so I bought myself a little book with short stories so I could put to fullest use my dictionary, as well as a Polish-English dictionary for fifty cents that explained much about the English language. Also, I soon found a Spanish sailor who was happy to help me, especially in the pronunciation of words. These three books and my infrequent encounters with my Spanish friend wholly occupied me.

On the third or fourth day on the ship, many of the passengers fell ill. I, fortunately, was not among them, but I had no appetite for any of the

## The Oath

prepared food. I got by on vegetables that I bought from the ship's store. I drank water with oranges and lemons to keep me free of any gastric problems, but that concoction was bad for my appetite. Demyan, however, did become very ill—he even thought he was going to die—but he received no medical attention. I fed him orange juice for several days, and he became better and soon was back on his feet.

On the eleventh day after leaving Porto, we arrived at around three a.m. at the port of Veracruz, Mexico, where we would dock for two days. I asked one of the ship's officers if I could take a little tour of the city. At first unwilling, he eventually gave me a boarding pass that would allow me, after leaving the ship, to return to it. He enjoined me not to leave for very long, as the counting of the passengers would take place the next day.

I approached the bridge, where a sailor scanned my pass, lifted up the chain that blocked the passage to and from the ship, and let me depart the ship. The city of Veracruz—crowded, dirty, and disordered—immediately made a bad impression on me. People in wide hats, some with blankets on their shoulders, moved like moribund flies in boiling water. All of them—without interest in life, careless, and uninteresting—looked as if they had been stunned.

I went to a pharmacy that had the German name of the owner on the sign. I asked the young gal behind the counter if I could meet the owner.

"Señor Palio!" she, without moving at all, called.

Behind the high cupboard, there came a middle-aged man with a handlebar mustache. I approached him and asked, "Do you speak German?"

"*Ja*," he answered.

"I want to ask you something about—"

"Hmm," he interrupted. "Please tell me—how does a fugitive from Poland, without money and who doesn't know the language, get by?"

"You are from Poland?"

"Yes, I'm from Poland. Do you know where Lodz is?"

"I know."

"Well, it's best for us to speak Polish so that nobody will understand us," he said in Polish.

**Picture 34.1. Havana, Cuba (1925)**

We struck up a conversation. The owner was a Polish Jew who had been a pharmacist since 1920. From him, I learned that living conditions in Mexico were bad, that there was no work, and that without a large amount of money, there was not much hope of making it. The climate was hot and humid. For people like us from the northern countries, the climate was unbearable. When he first moved to Veracruz, he said, he could not accommodate himself to the heat. He drowned in his own sweat for whole days and nights, and his body was always covered with a rash.

The sentiments disturbed me. "What should I do?" I asked. "I have a visa for Mexico and for Cuba. Is Cuba better?"

"No, not better, but Cuba's an island, and so the living conditions are easier, better. Well, it's not far from America. If they change their laws, you can always go to America."

He treated me with some sort of water called cassia,[78] which seemed to me pleasant in smell and taste. I thanked him for his hospitality and went back to the ship.

I had some trouble boarding the ship. Another sailor was on the bridge and guarding the entrance. Muttering something in Spanish under his breath, he rotated my boarding pass in his hands to all sides, but he did not let me in. I asked him to call the sergeant who issued me the pass, and he would acknowledge that I belonged on the ship. He looked again at my eyes and looked at the pass and finally unleashed the chain, and I entered the ship.

I met Demyan on the deck. He was happy to see me, though a bit angry. "Where have you been? I've been searching the whole ship for you!"

"I was checking out the city."

"Well, it's—it's nothing. I just didn't know where you were."

"Well, you don't need to know everything," I answered with a smile. "But come here—come here. Let me tell you something."

We found a spot where we were alone. I then told him that I was in the city and had gained some information from a German pharmacist. I told

Demyan what the German told me and that, consequently, I had decided not to remain in Mexico, but to go to Cuba. Though naturally antagonistic, he did not protest.

From Veracruz, we sailed to the small town of Campeche, where we docked for several hours before heading toward Cuba.

On May 10, 1924, we arrived in Havana. We quickly learned of the need for pocket money. The Cuban government only accepted passengers who paid at least fifty dollars to step on Cuban soil. Otherwise, the Cuban government would not accept any immigrant.

When we left the ship and found ourselves on Cuban soil, we chanced upon a well-dressed man who spoke Ukrainian. It made us feel dizzy with happiness.

He was a Galician Jew, and he owned a tavern and had a few unoccupied rooms on the first floor of Sol Street for twenty-five cents per day—seven pesos per month. The price was right, and the fellow spoke Ukrainian, so we decided to stay there.

We settled into our room. We bought food in restaurants and in the marketplace. We wrote letters, for the first time, both to America and to Ukraine, to let our relatives know where we were. We told no one to write back to us because we did not know how long we would be where we were.

We passed our time in Vedado Park or at the pier. We visited the U.S. consulate to inquire about a visa to America, but that was a waste of time because there were thousands of people who wanted visas to America and only a few visas. Some were waiting for two years. It was also impossible to find work, for there were few jobs available and for each job there were hundreds of hands waiting for it. Moreover, the pay was paltry.

# The Oath

There was a possibility of work on the docks at the wharf, but there was the threat of a strike. My situation was unenviable. I did not see any way at the time of repaying the large debts my father accumulated to send me abroad.

While sitting in Vedado Park one day, we learned that a certain Joseph Dycky of Canada was there and was recruiting workers. He would be in the courtyard of the Baran tailor that same evening.

In the evening, we came to Baran, a native Ukrainian from Sambirshchyna. I was astonished that there were already plenty of people like me in the courtyard. I would later learn that there were some two thousand Ukrainians in Cuba, and many were worse off than was I.

At around nine p.m., Dycky arrived and was greeted with loud applause as our savior. Dycky, who called himself Doctor Dycky, made a good impression on me. His Ukrainian was clean, and he was very elegantly dressed in Cuban fashion, in white clothes. He said that he understood our misfortunes and that he came to help us. He promised to take all of us to Canada, where there was much work for everyone.

Dycky was heartily applauded. He then said that all willing to go to Canada must sign into his book. To ready our trip, he required each "worker" to pay five dollars of pledge money to Mr. Baran, who would enter the pledge in the book. That money, he said, would ready documents for each of us and a first paycheck, yet everything was to be subsidized by the Canadian government.

Caught up in the excitement and swept away by optimism, each person paid the five dollars. For many, it was their last five dollars, but so great was the desperation and the desire to escape the hopeless hell that everyone happily paid. Dycky gathered about three thousand dollars, and we would never again see him. He and Baran were swindlers, and they seriously insulted and outraged the luckless poor people, perhaps his own people, on the tropical island.

# Chapter 35

## "You're here looking for me!"

HAVANA WAS A VERY BEAUTIFUL and cosmopolitan tropical city. Built around a bay of the same name, it had many parks with various tropical and subtropical trees and plants. Wide asphalt boulevards in the middle of the city and beautiful multistory stone houses surrounded by white and gray marble gave it a fairy-tale beauty. The city also had a light industry, a large factory of meat products, and a number of oil refineries. Stores and especially taverns did not have a front wall, but instead a rough steel wire, which could be secured at night and opened during the day to allow everyone free entrance and exit. In the middle of the city, there were good tramway carriages, numerous buses, and taxis at every turn.

The whole industry, including the large warehouses and luxury stores, was in the hands of American Jews, yet that was not apparent at first glance. Yet when one looked closely, one could see that native Cubans were not running any of the businesses. They were, instead, the laborers. Still, there were many European laborers because native Cubans were lazy, uneager to work, and uneducated, and much of the work required some degree of education.

There was a state lottery in Cuba, and each store, even one dealer, sold a lottery ticket for twenty-five centavos. The lottery brought much money to Cuba and to the large and small traders. There were winners each week. That lured many native Cubans to play, and some would spend all of their money with the hope of winning.

## The Oath

There were many dilapidated buildings in Havana, and many of those buildings needed to be disassembled so that modern skyscrapers could be built. Many of our people found work in disassembling those buildings because that work needed to be done by human hands. One of our emigrants boasted to me that in three months he had earned ninety dollars disassembling buildings, and that was a large sum of money, compared to what he earned in Poland—"great prosperity," as he called it. I was eager for work, but I refrained from that type of work, which was exhausting and required strength that I did not possess.

The food in Havana was inexpensive. An unpretentious man could live for twenty-five centavos a day. But when there was no work, then whatever money a person had would not last long.

On one rainy day, Demyan and I were sitting in our hotel bar. It was hot and sultry in the yard—another of many hot and sultry days—and we, like all others, were sick of the heat. Next to my table, there sat a man who betrayed through his mannerisms that he was not a Cuban. He ordered a beer.

"Pole?" he asked while sipping his beer.

"No, German," I replied.

"Do you understand Polish?"

"No," I lied.

The man tried to speak Spanish, but he knew that language worse than me.

I thought that he was not interested me, but for some reason, he could not take his eyes off me. As all the buttons on his jacket were buttoned, he unbuttoned the jacket to get a scarf, and I saw a familiar plaque with the eagle and the inscription: Polish State Police Intelligence.

*Aha!* I thought. *You're here looking for me!*

Poland did not have a representative in Cuba. Her affairs were at the Romanian consulate. This man did not know me by appearance, but if someone could identify me by, say, pointing a finger at me, he could call the Cuban police, and that would help him finish his devilish deed. I knew that a few fugitives from Poland had already been caught. I was glad that I had been relatively quiet about being Ukrainian, as no one could betray me in my native language. I got up and walked toward the bathroom, then took the side door that went out to the street and got lost in the crowd. That is how I lost the Polish policeman, who without question was looking for me.

# Chapter 36

## "Somehow, the Lord will help us"

**Picture 36.1. Boat at Sea (1925)**

ONE DAY, AS WE sat near a harbor in a crowd, a well-dressed man approached us and asked if any of us wanted to go to Cárdenas, which lies eighty-five kilometers east of Havana, where you could get better work on plantations than in Havana. Several of us agreed. The ride would cost two pesos and would happen by a large boat.

# The Oath

It was arranged that, on the second day, we would meet the man near the pier. He asked for no money from us in advance and gave everyone who agreed to go his card.

When I returned to my room in the evening, I saw that I had had uninvited guests. My suitcase was gone, and some of my clothes had been taken. I was glad that I had taken my jacket, in which I had sewn some two hundred dollars, with me. I reported the theft to the hotel's owner, but he merely said that he had not seen anyone go into anyone else's room. I complained that I would call the police, but I did not because the damage was not so great and I did not need any attention drawn to my affairs.

Before five a.m., I and five others were at the pier. Within a few minutes, a large boat with about thirty others in it arrived, and we boarded it. There were women with young children and men from every country in the world, it seemed. There were three Ukrainians: old Yatsina, Demyan, and me. The owner of the boat, Señor Paul, took from each of us our two pesos, gave us a ticket, and then the boat moved westward from the harbor gulf and spun sharply to the north. In an effort to avoid the savannah swamps, it moved far into the Mexican Gulf. Señor Paul had two mates. One was sitting near the motor, and the other was near the stern. Paul stood on the roof and looked to the north. The expansive sails of the boat made a shadow.

"When will we be in Cárdenas, Señor Paul?" I said.

"Oh, in about an hour," he answered calmly.

At some point, Cuba was no longer visible, and it seemed to us that we were not in the Gulf but in the middle of the ocean. The sun, though near the horizon, beat mercilessly on us, but there was nowhere to hide. We were entertained by a large number of dolphins that swam near the boat. They popped up high into the air and dipped into the water again. Often swimming under the boat, they seemed to be playing with us. I wondered whether one of the dolphins could tip the boat if it struck it firmly. At some point, something seemed to push the boat. Señor Paul rushed downstairs. He was there for a long time, maybe an hour. I leaned over the stairs to see what was happening there. He and the mechanic stood over the motor, which had overheated. They were arguing about something. I realized that something was wrong with the motor. I went to the rear of the boat and saw that the propeller did not work. When Paul came out to wipe his sweaty forehead, he looked angry.

"What happened, Señor Paul?" I asked with a soft tone.

"The jack snapped, but don't—don't worry. We'll fix it. We'll fix it—somehow."

There was no wind and so the sails were unavailing.

The repairing went on into the evening. In the evening, when the sun went far beyond the horizon, we were tired of the heat, so we stretched out on the deck and fell asleep.

We were awakened in the morning by a warm sun. Dirty from sweating the prior day, there was no way for us to wash. The barrel of water in the store of the ship was exclusively for drinking, yet it was spoiled. Moreover, there was no food. It was frustrating to sit—each of us was thirsty, hungry, and dirty—and merely wait till the engine was fixed or the winds were appropriate for sailing. Yet there was nothing else to do.

At around noon, Paul and his assistants brought us two large cans of beans. We had not eaten in more than a day, so even canned beans proved tasty, yet the canned food merely intensified our thirst. Later, Paul gave each of us a small glass of rum, which lessened somewhat our thirst.

We did not know where we were. We did not know whether we were floating or standing in place. Even the dolphins that sometimes played around no longer amused us.

# The Oath

The blazing sun beat down on us and redoubled our thirst. The water in the barrel was now off limits. Filled with dead and rotting insects of all sorts, it was unfit for consumption.

"If only a wind would turn," said old Yatsina to me. "If only—if only a sip of water—"

"While a sip would slake your thirst a bit, the water—it's very unsafe!"

"What do I care? If only I could be cooled a bit! That sun—well, it—it's scorching us!"

"Be patient, Mr. Yatsina. Somehow the Lord will help us," I replied.

The evening again cooled us. Thirsty and hungry, we fell asleep with no one moving around much.

On the third day, no one got up, because there was no reason for getting up. The young Armenian woman, who had with her a baby of some seven or eight months, pulled her baby, who had become ill, to her breast. The poor little child moaned and cried but would not take his mother's milk. Near her was a black man of about thirty-five years of age who looked helplessly at the mother and baby. We sympathized with the mother and baby, but no one could do more. She and her husband spoke nothing but Armenian.

I asked Paul if he could help the woman in any way. He merely shook hands with the mother.

Looking at her dying baby, the poor mother's cheeks were filled with her tears. It was not difficult to guess what was happening in her tormented soul. It was a sorrowful tragedy, and all of us were in some sense implicated in the fate of the baby. Yet each of us began to worry that the baby's fate would also be ours. We were enveloped by fear and infuriated and, most of all, tormented by our thirst and the lack of water.

The water in the barrel, full of many different dead insects, had a foul smell. When someone wanted to refresh his lips with water, he covered his nose and mouth with a napkin so that he would not smell the fetid odor through it, then put his napkin-covered mouth to the water to keep the worms and insects from his mouth.

The passengers' anger and fear began to manifest itself through grumbling and unrest. Yet Paul and his assistants always carried revolvers in the event of danger aboard. At some juncture, the assistants came out with rifles. It was evident that they were afraid of rebellion and were ready to shoot anyone who might turn a bad situation into a rebellion.

German Fritz Wraugger, who was very attached to me and very faithful, prayed often—especially at night, when everyone was asleep. His piety influenced many on the boat, who did, in their own manner, what Fritz did. In prayer with Fritz, we found some relief of our plight and some hope for salvation.

In the evening, the Armenian child died. The poor mother, holding the corpse to her chest, cried uncontrollably. None of us could console her, because she did not understand us, yet her husband fared no better. After a while, she placed the dead baby on the boat. She then leaned on her husband's shoulder and, shouting some incomprehensible words, began to sob loudly.

We all looked at her with deep sympathy. I approached Paul and asked him to give her something to drink to calm her. He went downstairs, took a few bottles of rum, and gave a large glass to the woman and her husband and a small glass to each of the rest of us. The drink slaked our thirst, but it exacerbated our hunger. Realizing that, Paul brought up tin cans of fish and distributed one to every two people. I have always disliked fish—even now I wholly eschew it—so I gave my part to Fritz.

After eating, Paul approached the Armenian, pointing his finger at the dead child and then at the sea. She said nothing. With pitiful eyes, she

## The Oath

looked to him as if to communicate this: "Do what you want, because you are the cause of his[79] death!"

Paul took from her the shawl with which she had covered the baby, wrapped up the small body, and went to the hole of the ship, where he cut a piece of cord and grabbed a bag. He placed the child's body in the bag, tied it, and returned to the deck. On the deck, he took off his large hat from his head, crossed himself, and made the sign of the cross over the bag. He then handed the bag to his father.

The baby's father remained stoic throughout the ordeal. He carried the bagged body to the side of the ship and, with the respect and ceremony only a father could give, dropped the dead body of the baby into the water. The man's wife, who was motionless, rushed like a wounded lioness to the side of the ship to be by her child. She roared with defiance and rebellion. Her husband restrained her, but she escaped his grasp and ran to the low rail on the side of the ship. She would have fallen into the water had not Paul turned and caught her in his strong hands. Several of us rushed to her from both sides and tried to calm her down. That was no small task. She scratched and clawed us, bit our arms, and growled like she was insane. Paul threatened to punch her. In the end, we sat her down and calmed her. The childless mother, now sitting, thereafter looked daggers at Paul, who was holding a rifle.

This tragedy of the unfortunate Armenian family touched the very depths of the soul of each of us.

Throughout this ordeal, we were very astonished we did not come across another ship, especially a fishing boat, which could help us. We could also not understand why Paul and his assistants were carrying weapons, but nobody wanted to ask them about that.

# Chapter 37

## "Lord, please keep us from the storm!"

AS THE SUN BEGAN TO set, there appeared small white clouds in the sky, which Paul examined. The sun would not set cleanly that night but would hide behind the clouds.

"Well, looks like a change of weather," I said to Fritz.

"Lord, please keep us from the storm," he said to no one in particular.

At twilight, the clouds thickened and merged. In the tropical and subtropical areas, storms came quickly, without warning. Sometimes they would come, disappear for some time, and then reappear. Leaning his shoulder on the mast, Paul continued to follow the thickening clouds with great attention.

"What does he—what is he waiting for? What does he know that we do not know?" I said to Fritz.

"Yes, really! I'd just like to know how this nightmare will play out," he replied.

When the sun set, with its setting there came a very terrible night. Fritz went to his knees to pray. I, too, went to my knees to pray. Each of us, I suspect—in his own language and in his own way—did the same. There would be misery and evil ahead, and all sensed that. No one would sleep that night.

# The Oath

A strong jet of warm wind from above turned our boat some forty-five degrees. Large waves formed in the once-tranquil sea. The rocking boat was being tossed around like so many shells of peanuts. A large dolphin left from the ocean and plunged back into it, as if to warn us of impending dangers. Paul and his two helpers took down the sails.

There was pitch darkness around us. A great wave of water flew out onto the deck and almost tossed us from the boat. We panicked. Believers prayed; others cried. There was no place to hide because there was no place for us on the deck. The women who were on the boat began to scream. It is impossible to describe the horror.

Beaten by raging waves, we were soaked with water. Paul ordered everyone to lie down and hold on to the railings, as that was the safest position to assume. Still, the waves fell on the deck and pushed us here and then there or merely beat down on us as if we were dumplings in boiling water. The water and wind created a deafening noise. Conversation was impossible. With the engine inoperative, the boat was without a light. In the middle of the deck, a small oil lamp shone, but its yellow light was not visible beyond a few feet. The wind swirled viciously. No one could think of anything but death. Still, each of us clung greedily to life.

Standing at the stairs, Paul was drenched. The yellow light of the lamp made his tall, strong figure seem nefarious, mysterious. I crawled to him and asked him to take below the five women. He agreed. Fritz and I dragged down one woman. Others brought down another one. We slowly brought down the women so each could sit on some sort of bench.

At low light, we saw that the boat was loaded with some boxes, one of which had on it "RUM." We then realized why Paul and his assistants were carrying weapons and why we met with no other ships or fishing boats. Paul was smuggling alcoholic beverages to America, where alcohol was prohibited. Such a business promised large revenues, and only a skilled sailor with full acquaintance with American laws and customs could pull off that. Paul, with compass and a map, deliberately avoided all other boats, especially the American Coast Guard, by sailing all the time in neutral waters. That was why he panicked much with the breakdown of the motor. He could no longer control his position at sea and might fall into the hands of the Coast Guard. That was also why he had scanned the waters with binoculars the past three days. This came unexpectedly to me, for he was a tall man with a strong body structure, a handsome face, and decisive behavior and gave every indication that he was an international merchant, not a bandit.

**Picture 37.1. A Florida Swamp of the Sort Chemny, Demyan, and Fritz Trudged**

## The Oath

Then came the rain, which proved a mixed blessing. Thrust at us by violent winds, the drops struck us like innumerable stings of a bee. Yet many of us opened our mouth to take in whatever we could of the sweet rainwater, as four days without water had taken a toll on us. Flashes of lightning lit up the sky, and violent thunder crashed on the horizon. The violence of both was such that no one on the boat could doubt that there was a God in the world and that, in this terrible hour, all we could do was to ask him for salvation.

Old Yatsina and young Demyan, a boy of weak will, were kneeling and praying in the face of the fierce wind. "We offer up our fear to you, God!"

The wind lessened, though it continued to rain and the boat was still rocked by large waves. No one could sleep.

In the early morning, at about three a.m., the rain stopped. There were no longer any large waves, and the boat stopped rocking.

As the sun was about to rise in the east, the clouds disappeared and we could see stars in the sky. We had survived God's wrath.

The sky near the rising sun was purplish, and a thin golden strip covered the sea. To my right, maybe a kilometer or two in the distance, I saw something black that I did not see to my left. That could only be land.

I tugged at Fritz's sleeve and whispered to him, "Look!" I pointed out the black strip.

"*Ja*," he said. "Land!"

"I am jumping into the water. What happens happens!" I said.

"I also!" snapped Fritz.

Demyan, who was lying beside me and had been asleep, woke and whispered, "Let's go!"

We slunk to the side of the boat, crossed the railing, and jumped into the water.

I thought it would be shallow where we were, but it was not—I could not reach the bottom with my feet—and so we had to swim for a while. As we got some distance from the boat, I again tried to reach the bottom with my feet. I was still unsuccessful.

One of the passengers noticed that we had left the boat. Soon, everyone was talking animatedly about our departure. The chatter waked Paul. With a rifle in his hands, he shouted for us to return. If we refused, he would shoot. Being some one hundred meters from the boat, we ignored his shouts. We merely made every effort to get to the shore sooner.

The coast, which had seemed so close from the boat, did not seem to get closer as we swam. Yet we continued to swim. What choice did we have? And so, we swam our way to a shoal where we could now walk even though we were still in water up to our necks. The water there was perceptibly warmer.

We continued to walk toward the shore. The sun began to rise.

We reached a swamp filled with high sedge. Tripping over the roots of the sedge and the various trees in the water, we trudged through the swamp for what seemed like an eternity.

# PART V
# A LAND UNKNOWN

# Chapter 38

## "This Isn't Havana"

WE WORKED OUR WAY out of the swamp, which was overgrown with sedges and bushes. That was an ordeal. This sedge, which I later learned was called saber grass, had very sharp teeth on its large blades. As we trudged through the swamp, it cut our pants up to our knees so that only threads of cotton hung. Worse, it devastated our feet, which were bleeding when we left the swamp. We were fortunate not to have encountered the poisonous moccasin snakes, which were numerous in that area and which hid between water trees and hunted for food in the morning.

Though completely exhausted, we walked on dry land. Following the rising sun, we walked all the time to the east. In two or three kilometers of our land journey, we found some kind of pit with clean and cold water. Having no cup or ladle, Demyan and I cupped our hands and scooped out and drank handfuls of the water. Fritz, who had never used cupped hands to drink water, watched in amazement, and then he did the same. None of us, it seemed, could take in enough water. It was cool and sweet—delicious. Never before and never since has water tasted so deliciously.

Having drunk enough, we followed a road to the east in the hope of finding someone who could help us. However, we could not walk far. In no time, we walked off the water. We were again thirsty—thirsty, weak, and exhausted. We became sleepy, and each of us felt acute pains in our stomach.

Not far off, we found a thickly wooded area. The land there was completely dry. The storm that we suffered at sea had apparently not reached this land. We lay under one of the trees and quickly fell sleep.

When we woke up, Fritz was gone. He probably had decided to make his way on his own and did not wish to wake us before leaving. I would never again see Fritz.

It was about three p.m. The day, we had come to find, was June 14, 1924. Demyan and I traveled east by the forest. The flora somehow seemed different. I saw no palm trees. Yet the stench and heat were the same. We were never in a Cuban forest, and so this was very different to me.

We walked for the better part of an hour. The forest was thinning. We saw large lawns and hayfields. In the distance, we saw two people putting hay on a cart harnessed to two beautiful black horses. We approached them with the hope of discovering just where we were and to see if we could get some food.

When we approached them, they stopped working and studied us. There was, of course, every reason for the men to stare incredulously at us. Our shirts were filthy. Our hair was dirty and uncombed, and we had no hats. Our pants, shredded below the knees, were filthy, too. On our legs, there was dried blood aplenty.

They were not Cuban—that I knew—but I could not determine their heritage. So I greeted them in Spanish insofar as I could communicate my sentiments.

"Havana? We—we Havana? Where?"

"Havana?" said the older of the two with amazement. Knowing of my limited ability with Spanish, he spoke slowly. "This isn't Havana. It's Florida!"

Demyan and I looked at each other. I composed myself and said, "Señor, food? Food? We—we very hungry."

"There's nothing nearby. The closest thing is Vyborg City (Вийбор Ситн).[80] That's about seven miles. But you can't go to the city looking like that! Come—come with me to the farm."

He threw up a few more forks of hay on a cart and told his son, a boy of some fifteen years, to drive us to the house. On the way, we passed an inquisitive Spaniard, who asked us who we were and how we got there. He spoke a mixture of Spanish and English, but we got the gist of what he said. I answered his questions as best I could. He appeared to me to be a good-natured person who sincerely sympathized with us.

When we arrived at a well laid out farm, we asked the owner, a certain Don Juan Gonzalez, for water so we could wash. We washed ourselves. After introducing us to his wife, he took us upstairs, where there were several rooms and one bed that nobody used.

"Here, use the room as your own," he said, "but first we'll get you something to eat."

His wife cooked something for us to eat, and the farmer asked us about our journey and whether we had anyone in America to help us. He spoke sincerely and with kindness.

I answered all his questions. I added that I had relatives in Detroit, my father-in-law, and his brother in Cleveland. I even gave him addresses. Demyan said that he had a brother in Cleveland.

When we had a little snack, we went to our bedroom, where our bed had been well laid out and draped over chairs were two old but clean trousers. Don Juan came to the room for a moment. He had with him writing paper and envelopes. "Here, you can write letters to your relatives—let

## The Oath

them know where you are, how you are doing, and so on. I will write the return address myself. When you are done, give me your letters and I'll take them to the post office."

We thanked him for his kindness and his hospitality.

"Don't forget the—the pants. You need them."

We began our letters. I wrote a short letter to my father-in-law in Detroit. I told him that I was in Florida. I then asked him to send me some clothes because I had none.

On the second day, most of the wounds on my legs began to heal, though some began to redden and fill with puss. The good man, Don Juan, brought us iodine to fight the infections. The iodine burned but it helped, and soon all of our wounds began to heal.

On the third day, we went out to the courtyard and helped the farmer with his work. He accepted our help with gratitude. I wished to show him that we were not parasites who fed on his hospitality.

We eagerly waited for answers to our letters. None was forthcoming. I began to worry that our letters never found their goal. Yet there was nothing to do but wait and work.

On the fifth day, a Saturday morning, a taxi arrived at Don Juan's farm. A man with a suitcase got out.[81]

When the mysterious visitor entered the house of Don Juan, I immediately recognized him: he was my dear father-in-law. Instead of sending clothes and money, he came down to Florida and brought us everything that was necessary, even boots.

Demyan and I dressed. My father-in-law gave twenty dollars to Don Juan for his kindness and brotherliness toward us. We said goodbye to our benefactor and left him.

We ordered a taxi to the railway station for a train to Detroit, which would depart at noon. My father-in-law bought us tickets, and we sat down in an ordinary train car, which—though ordinary—was better and more elegant than a Polish or a French first-class car.

On Sunday morning, we were already in Detroit. One of Demyan's brothers was there to take him to Cleveland. He left and we parted. I settled in with my father-in-law.

At my father-in-law's house, I washed, and we went to the Ukrainian Catholic Church on Grayling Street. I prayed wholeheartedly to Almighty God for his kindness, which had saved me from apparent death and brought me to this country in a healthy state.

After mass, we entered the Ukrainian bazaar, where I met several active Ukrainians. On the same day, I wrote a letter to my wife and relatives to tell them about my arrival to America. I assured my father that I would work—even perform the worst work—so that I could earn enough money to pay off his loans.

My father-in-law worked the night shift at Maxwell's factory, so he had his days free. He drove me in his car around the city and to the factories where I was looking for work, which was not easy to find. Wherever we went, there was no work. I was worried about disappointing my father-in-law. I did not wish to be a parasite and live off his kindness.

The days passed. I went from factory to factory, but nobody hired me. My days were thus filled with sorrow.

One day, I met near one factory two Ukrainians from Canada who were

# The Oath

looking for work, too, and like me, could not find it. They told me that the Detroit Edison Company's Detroit Power Plant was building a new power station near Detroit and needed workers there. I told that to my father-in-law. He was not so excited. It would be hard work for me, he said, but he gave me a few dollars, and on the road I went.

I was offered a job for thirty cents per hour, and I accepted. I worked there for ten weeks. I shoveled cobbles and sand and mixed cement with crushed stone and sand. It was hard work, but I was glad that I had work. The experience was, overall, positive.

It was already autumn, and the mornings were frosty. The other workers and I slept in the barn of one farmer and took our meals there. For food and "lodging," the farmer took half of our salary, so there was little to gain by working at the power station.

The barn was impossible in the winter, so I quit the job and returned to the city. Nevertheless, I saved so much that I bought a winter raincoat and sent my father fifty dollars.

Back in Detroit, I found work in a factory where I wound copper pipes. I enjoyed the work, which was honest; I got forty cents per hour; and no one took half of my paycheck. Yet the work was dirty and wet because the hot pipes were laid in an iron trough that was filled with water and some oil, and from there, the pipes went to a special device that forged them into a specified thickness. It was cold in the factory in winter because there was no heating. The only warmth to be found was in the trough where the hot pipes were wetted and cooled with warm water, and one working the trough could get bathed in hot steam to stave off the cold.

I worked there throughout the winter.

In the spring, I suffered an accident. One of the thick-caliber pipes broke during a stretching and tore open my hand, and I could no longer work.

No factories offered employees health care at that time. Everyone who injured himself at work was responsible for doctoring himself at his own expense. Even if he could manage to see a doctor, he would have to pay for his own medicine or his own bandages.

I passed the winter by going to school in the evenings to learn English. In March 1925, I got my diploma.

When my hand was healed, I got a job at a spring factory. The work was lighter, and it paid more, but the factory was very hot, as it was necessary to stand near the heated furnaces and by a trough with oil where the springs were rolling.

# Chapter 39

## Settling in Detroit

**Picture 39.1. The Beautiful Immaculate Conception Ukrainian Catholic Church in Hamtramck**

DURING THAT TIME, I met some of the Ukrainians living in Detroit and around the neighborhood. Through them, I became a parishioner of the Church of the Immaculate Conception of the Blessed Virgin Mary in Hamtramck, signed up to the patriotic Ukrainian organization Sich, and became a friend of Father Dmytro Dobrotvyr. All the while, I was chary of freeing my dear father from the debts he incurred on my behalf.

Detroit was not Khom'yakivka. At the annual parish meeting, I felt very like a stranger because parish officials acted with more pomposity than

## The Oath

did the village heads that I had left in Ukraine. The parish's priest was treated as a servant, not as a priest. To such disrespectful behavior, I was unaccustomed. In Ukraine, the priest ran the parish, not a church committee consisting of semiliterate people who made wild ordinances, as did village youths.

I offer two examples of such wild ordinances. First, there was a rule that no one could come to the church to pray unless he was recorded in the register of parishioners. I related to the committee members that the church was a house of God, and nobody had the right to forbid anyone to come to pray. Second, there was a rule that if the church was full and a stranger had a place in a pew in which to sit and a parishioner did not, then the stranger could be expelled from the church. I told the committee members that this was a silly rule—one opposed to the principles of Christian benevolence. They replied that they built the church and so they could make the rules.

Yet as members of Sich, the same people behaved exemplarily. They listened to their *otaman* without any bias, acted with dignity, and treated their elders with respect.

Two weeks before my arrival in America, Sich, at the Fifth Congress of Philadelphia, adopted a hetman ideology and adopted the motto: Cultivate discipline, respect secular and spiritual authorities, and defend the historical truths of Ukraine.

While Dr. O. Nazaruk wrote intelligently on those subjects, there were many rebellious democrats in America. They refused to follow the Ukrainian hetman Pavlo Skoropadsky. Against the Sich, there arose the magazine *Svoboda* (Свобода or *Freedom*) and its radical movement. With its editors, Dr. Luka Myshuga and a certain Reviyk, there began a bitter political struggle with large implications for Ukrainian Americans.

In the summer of 1925, Bishop Constantine Bohachevsky arrived in America. The two editors of *Svoboda* viciously and for no good reasons attacked him, and in addition to the sociopolitical struggle among conservative and radical Ukrainians, there began a church struggle. In defense of the bishop, there was Dr. Nazaruk. The fight was bitter and no good would come from it.

The arguments by conservatives were clear and cogent, but they proved unavailing, as *Svoboda* assailed its readers with clever lines and quixotic notions to turn Ukrainians against the Skoropadskyians[82] and against the bishop. As a consequence, *Sich* magazine declined in popularity. Some churches dropped out of the diocese and became Orthodox. Others remained "Catholic" but turned their backs against the bishop and against those who still held their great-grandfather's sort of Catholicism. The bishop, who had only recently been released from a Polish prison—imprisoned only for being a dyed-in-the-wool Ukrainian patriot—had arrived in America only to be called "Polish servant," "Pole," "Roman tutor," and so on.

To work constructively in bitter circumstances and to keep our church together, we had to have iron nerves and forget petty squabbling among ourselves. We were fortunately up to the task, as we parried courageously undeserved slander and savagely fought the evil of radicalism.

The "Hetman Sich," so-called after acceptance of hetman ideology, had to become a thick brick wall—the bishop being the cement that bound the bricks—to block all kinds of attacks from the mercurial people who were incited by *Svoboda*. Each Sichman had a duty to defend honorably and proudly Hetman Pavlo Skoropadsky and Bishop Constantine Bohachevsky. They fulfilled well and with dignity that task, and it can be asserted that, where there were Hetmanites, there was no chaos. Thanks to the Sichmen, our church was not riven by dissention.

The writings of Dr. Nazaruk had a grip on me because I was used to order, and radicalism was strange to me—strange and insufferable. I soon became a member of Sich. I dove with all my energy into social and public issues concerning Ukrainians and Ukrainian Americans. After several contributions to *Sich* the magazine—several essays and one poem—Chief Otaman, Dr. Stephan Hrynevetsky, nominated me as a cultural and educational referent, with the rank of an otaman.[83]

Things were not easy for me at work, which was very physical. My mentor, a hot Irish patriot, returned to Ireland, and his position was given to a man of little education. One late spring day, when I was sitting around a boiler where the steel for the springs was, he drew from the oven an almost white-hot tin box containing springs. Instead of turning to the right, he turned to the left, and the white-hot tin box rubbed against and burned my whole arm below the elbow—especially my hand. The burn was terrible, and my hand was smoking. I was taken to a doctor, who smeared the wound with some kind of ointment and placed my arm in a sling.

The pain did not diminish but intensified. In a few hours, I felt terrible burning pain in my whole arm. I could not sleep all night. So great was the pain that tears welled in my eyes. At some point, I asked myself why the Lord always punished me so. On the third day, the pain stopped, but the burn was covered with mucus. For a whole week, the doctor daily dressed the wound. For six weeks, I walked with a bandaged hand. Six weeks were thus wasted, as I got no money for convalescence; and there were, of course, the doctor's bills.

I passed productively the period of convalescence. I wrote articles for newspapers on a variety of social and political topics.

I was preparing to enter a technical university to finish my technical education. I also wrote to my father for him to send me my school certificates. My father replied that the Poles branded me a coward and had thus removed from the house everything that was associated with my name. In addition, he told me that my wife was being pestered. She was even called to the church to swear that she did not know where I was.

"We," wrote my father, "did not write to you because we knew you had your own troubles."

Consequently, I had to take oral and written examinations because I had nothing to certify my educational level. I passed both examinations and matriculated at the technical university.

When my hand was healed, I got a job at Packard Motor Company, where I worked at night, and went to school in the afternoon. I studied for five and a half hours, while I worked for eleven hours in the factory. I became so thin in the next six months that even my professor drew attention to it and told me that I had to choose between work and study. I did not have to choose. Packard soon eliminated the night shift.

Meanwhile, on the day of November 25, 1925, my mother-in-law arrived with two daughters, Hanna and Eugenia, from Europe. My father-in-law had to create living spaces for them, for his family was now larger by three.

# Chapter 40

## "God is not nonsense!"

NOT WORKING, I HAD to suspend my education. The cost was too exorbitant. So I wrote to my father's brother, who lived in Cleveland, to see if I could find work there with his assistance. My uncle never cared much for me, as he was pro-communist and an atheist. Still, he invited me to stay with him. He had asked his boss about a job for me, and his boss found me one. My uncle was proud of that.

As I had no special skills, they gave diverse tasks for me to do. That I liked. I did, without complaining, what I was asked to do, and I was appreciated because I did not complain. My salary was comparable to what I was earning in Detroit.

I came to an agreement with my uncle that I would pay him so much money per month for letting me live with him.

Working at the same place, we left together in the morning and returned together in the early evening. The trip was long because my uncle was in the west part of the city and the place of work was in the east of the city, and so we had plenty of time to talk each day.

We disagreed on many issues, but I merely kept quiet and seldom contradicted him. My uncle was not only an atheist but also a liberal cosmopolitan, whereas I was a Christian and Ukrainian conservative. Thus, our social-political views were antipodal, and we could not find a common language. He always thought he would turn me around to his way of thinking—not only because he was older than me and had more life

# The Oath

experience, but also because he had been in the Austrian army and he had lived longer in America. The problem was that he would never read anything other than Marx's *Das Capital*. I often asked him about various other writers on political economy, yet he only made big eyes and had nothing to say. He was, I knew, insecure. He was a proud man who conversed only with ignorant persons, but now his brother's son, more intelligent than him, had come to live with him. And so we could not have any sort of relationship. In the evenings after work, I read books or wrote, while he would go to sleep.

A week later, I enrolled at Lincoln University, and four times a week in the evening, I studied. On other days, I worked out my tasks. In February 1926, I wrote a patriotic poem, "To the People," which Dr. Nazaruk, who much praised the poem, published on the editorial page of *Sich*.

My uncle saw the poem, came to my room with an ironic smile, and said, "If you are such a poet, could you compose something more intelligent?"

"Is the poem so stupidly written?"

"It's not stupid, but it's not—it's not very wise," he said.

"Are you such a viable critic of literature?" I, offended, replied.

He left my room and went to sleep.

Angry and staring blankly at my books, I sat for some time, Finally, I, too, decided to sleep.

At breakfast on the next day, my uncle did not speak one word to me. He was also silent on the ride to work. He was offended at my remarks. Yet I, too, was offended at his criticism of my poem, though I did not want to admit that.

We did not talk for almost a week. Yet such childish silence got the better of me, and so I decided to break the silence. I did, but he did not answer me. I then said, "Why are you angry with me?"

"I am not angry. I merely have nothing to say to you," he answered with feigned indifference.

"No, we have something to talk about, but you are afraid to converse with me. Tell me if I am bothering you and your family, then I can find another apartment."

"As you please. No one has bound you here, though I'm not throwing you out," he said without much moving his lips.

"I am sorry to say this, but you are walking a dangerous path. You will not get anything from the communists except disappointment. You live a certain way and cannot see what others see."

"And what do others see?"

I could not hold back. "One can see that there is no love in your house, no sincerity, no happiness. You are a stranger to your own family! The Lord has already punished you! Marusya, though a beautiful girl, is deaf and dumb. Gania, who is also beautiful and has finished high school—and one could consider her to be a good girl—well, she has no respect for you and your wife, as her parents. She goes where she wants and returns when she wants. Your wife is also on the verge of giving you a nervous breakdown or maybe, God forbid, worse! You do not have God in your house! You live a brutal life! That is what—that is what another person can see! Yet—yet I say such things because I am afraid. You are still dear to me."

"You do not have anything to fear. About God—well, I have had certain experiences and—well, I don't believe in such nonsense!"

"That is too bad! God is not nonsense. But you can still change your mind—it is not too late—otherwise I much fear for you! I cannot put words to such things."

"Why?"

"Because there *are* no words!"

"Hmm, you don't know. You have no words. No longer bother me with your advice and your—your fears. I'm older than you, and I understand much about life."

"You merely *seem* to understand life. There can be no understanding without belief in God. Well, then. You say that you understand life. Well then, why is it that life always flees from you? Why have you never felt happy—satisfied?"

"How do you know that?"

"I read it from your faces—see it in the family interactions."

"If you don't like this *unhappy* house, you can leave at any time!"

"No, that is not what—no, not what I mean. It is just that I am sorry for you."

"I thank you for you pity, but I can easily do without it."

"Please, don't be angry at me if I touched a nerve. I shall be quiet."

"Well, you don't have to!"

After that, I lived with my uncle not as a nephew but as a lodger.

In the spring, my uncle bought a car, but he could not drive it. I took him to the park, where I could teach him how to drive. Learning to drive was a proud moment in his life.

Easter was coming, but in my uncle's godless house, it was an ordinary day. There was no preparation, neither for the body nor for the soul, for the holiday.

On Good Friday, I was going to go to the church after work. I asked his daughter Hannah to go with me. She gladly agreed but added that she did not know how to pray and that she had never been to a church.

"It's nothing!" I assured her. "Just watch me—watch what I do—and do the same!" (I would later describe this event in my short piece, "First Time in Church," which was published in *Philadelphia Way*).

When we returned home, both my uncle and his wife asked Hannah where she had been. She answered plainly that she had gone to the church with me. My uncle was furious that I would take his daughter to church without his permission, and cast some obscene words at me.

When the flood of obscenities ended, I, offended, said, "Well, I shall leave you tomorrow. But remember—remember that God may punish you for it!"

I went to my room, gathered my belongings, and then rested on my bed. In the morning, I woke before anyone else had awakened. I placed rent money on the table by my bed along with a note that read, "Be healthy! Let the Lord be merciful to you!"

# Chapter 41

## Reunited with My Family

**Picture 41.1. Packard Motor Company (Detroit) in Chemny's Day**

I HAD HAD ENOUGH of cleveland and my uncle. I purchased a ticket to Detroit so I could be with my wife's family and celebrate, not ignore, Christ's resurrection.

After the holidays, I got another job at Packard Motor Company. The work was easier and less stressful, as I had worked there before.

## The Oath

Returning to the university, however, was impossible because I had the afternoon shift and so I was free only in the evening. However, I soon learned that the Society of Polish Engineers was offering courses on practical engineering. I took two courses over two years and spent three hours of three days of the week in study. I finished the two courses. Yet when I was in my third year of study, I again lost my job, because the night shift was eliminated and the day-shift workers were moved to the afternoon shift. I thought, *He is an orphan for whom every wind blows in his eyes.*

I told Professor I. Shimanski about my "luck"—that I had to withdraw from school. He thought, and then said, "That's pity! You're a fine student, and we're sorry to lose you. I was hoping you could complete the courses and enter the company of Polish engineers."

"But Mr. Professor, I would not be accepted anyways."

"Why?"

"Because I'm not a Pole. I'm a Ukrainian!"

"What do you say? I always thought that you were a Pole. Hmm, that won't hurt you in the end. Why don't you go to Ford? They're not laying off—at least, not for a while—and there's technical work there."

"But, Professor, I do not know any technical work."

"Eh, such a big deal? Do you think all the 'specialists' are so wise and talented? In the end, if they lay off, they lay off. The crown will not fall from your head!"

I took Professor Shimanski's advice and went to the Ford Motor Company at Rouge and found a fairly technical job. I was assigned to the so-called Give Room (Дай-рум), where we produced all kinds of steel shapes to be cut and bent to fit various parts of cars.

For two weeks, I worked on a day shift and attended my third course and third year at the Polish technical school. Within a span of two weeks, however, I went from the morning shift (seven a.m. to three p.m.) to the afternoon shift (three p.m. to eleven p.m.), and finally to the night shift (eleven p.m. to seven a.m.). The shift to afternoons meant that I had to say goodbye to my professors and colleagues. Shimanski still was glad that I had gotten a good job with his intervention, and I thanked him for his assistance.

I performed my tasks well because I loved the work, which I found interesting, especially since it required some thinking. Precise measuring was part of the job.

Within a week, the director of our department asked me to come to his office. There, he asked me how much I earned per hour. He raised my rate of pay to one dollar and five cents per hour, and that was fairly high remuneration for work in those days. I, of course, was flattered and honored. I thanked him.

The change of shifts took a toll on my health. A bad diet during the war, irregular sleep during the arduous trip from Ukraine to America, and now sudden changes in work shifts caused stomach pains and a sore, burning throat. I fought those pains with different pills, but they took time to work and were mere palliatives, not curatives.

I worked hard and took off no time, and thus I was able to save some money. With money, I had an opportunity to bring my wife and daughter to America. I tried various measures, but all failed.

Early in 1927, I got the address of my friend E. Vasilyshyn in Canada and asked him if he would help me in this matter. He agreed to work out all the documents in my name and bring my wife and child to Canada. My father,

## The Oath

who had extensive contacts, helped my wife to get a passport and a Canadian visa. And so my wife and daughter arrived in Canada to stay with my dear friend Ivan Skaletsky in Windsor, just across the Detroit River.

Still, things were dicey because I was an illegal immigrant. Fortunately, I had joined the Michigan State National Guard in 1926 to testify to my loyalty to the United States of America. Yet, as an illegal resident, I dared not to leave the United States; but when my wife and daughter arrived to Windsor, I rushed via taxi to Windsor without taking any documents with me except for the card, which showed that I was a member of the National Guard of Michigan.

So excited was I to see my family that I thought little of the Canadian border guards. Still, I entered Canada without a glitch. I was there reunited with my wife and daughter. I remained for a few hours in Skaletsky's home to get reacquainted with my family and, I suspect, because I began to worry about reentry in the United States. I wanted to have some time to spend with my estranged family in the event that I, a fugitive, would be arrested and deported to Poland.

While I spoke to my wife and played with my daughter, I prayed. I asked God to allow me to return safely and without incident. My prayers were answered. Skaletsky drove us to the border, where he left us. When a U.S. border guard asked for identification, I submitted my National Guard card to him. He said nothing and let us return to the States. How I thanked God that night!

On Saturday, September 26, 1927, my wife, my daughter, and I arrived quietly in Detroit. We walked a little around the city, and then we took a taxi, which brought us to my father-in-law's home. The happiness was extraordinary. The family—comprising the parents and three daughters—was again together. I cannot express easily how I felt. My wife and daughter and I were finally freed of vicious Polish yoke.

**Picture 41.2. Fourth District Hetman Sich.** Central Secretary M. Melnykovich, Dr. O. Nazaruk, and Otaman Toma Koschil

For some time, we lived with my father-in-law. Having put aside some money, as I continued my job at Ford Motor Company, I purchased a house of my own.

My happiness was offset by one sad event. My great friend and teacher, Dr. O. Nazaruk, was leaving for Europe. He had me pledge that I would continue to work in the ranks of the Hetman Sich and, in general, for Sich ideals.

"How can I not?" said I. My eyes welled with tears. I knew too well that no man or group of men would replace him. Dr. Nazaruk was indefatigable and unselfish. Still, he promised to defend the Ukrainian Catholic Church and its ecclesiastical hierarchy, even from Ukraine. His loss was more than painful.

# The Oath

Our farewell was further aggravated by the fact that the religious struggle, started by the anarchists Dr. Luka Mishuga and Reviyk, was in full swing. There was great need to defend the truth. *Sich* thereafter had many editors until Dr. Mykola Bodruga assumed the editorship. Dr. Bodruga, like Dr. Nazaruk, was a staunch defender of the established ecclesiastical authority, but he did not have Nazaruk's command, vision, and energy. Mishuga and Reviyk also had a whole number of contributors to *Svoboda,* who often in an unscrupulous manner castigated both Hetmanites and bishops. It was then difficult for *Sich* to defend traditional Ukrainian religious and political values.

Consequently, Dr. Bodruga invited me to help the cause, and we made every effort to fight the drivel of the anarchists. Nonetheless, I was overextended. I was working at Ford to feed my family and completing my education, and so it was not so easy for me to write articles in order to rekindle our forgotten "democracy."[84] Yet I did what I could, and I contributed something, if only a small article, to every edition of *Sich*.

My efforts paid dividends. It seemed to me that Ukrainian Americans were slowly beginning to see the sense of order and structure in Ukraine and in Ukrainian America and to see through the lies of the Ukrainian radicals.

As a result of my work with *Sich*, however, I made many enemies who threatened me with beatings and even deportation. Yet my conscience was clean. I knew that I had chosen the right path, and so I ignored the threats. It was not easy to do what was right. Sometimes it was necessary to throw a bone, as it were, to our communist adversaries for the sake of ultimately disclosing their lies. It was very dangerous for me, as an illegal immigrant, to fight with the radicals. Some would have no scruples about inflicting the greatest injustice on me—deporting me. But somehow, the Lord guarded me from that.

# Chapter 42

## The Great Depression

THE YEARS 1928 AND 1929 were gray for the world, particularly for America, which suffered a grave and debilitating economic crisis—though I admit that, for me, they were not so dissimilar from prior years, all of which had been gray.[85]

In December 1928, I fell ill with pneumonia. For two weeks, I lay almost unconscious. Yet thanks to the sincere and skillful care of Dr. A. Kibzey, I got up from bed in four weeks and dragged my weakened legs to the Ford factory to ask when I could get back to work. I was told that I had been released from work because nobody told them about my illness. My wife, who had little command of the English language, could do nothing, and I, being almost unconscious, could not inform the doctor to contact my employer.

We were hoping to replenish our family, and there was I, without strength and with little hope. We asked the Lord for help, but that help was not forthcoming.

The next day, I appeared at the Baker Factory, where they gave me a very good salary and a good job, but it was less than ideal because I had to work at night. The shift, eleven hours, was too long, and it proved exhausting, but I endured for as long as I could. I at least had work.

On February 4, 1929, the Lord blessed us with a second daughter, whom we called Natalka. Yet we lived in a foreign house whose owner was a Pole and a wicked man who often fought with and beat his wife, and during

## The Oath

such rows, she frequently sought shelter with us. It proved impossible for me to bring up a family in such a volatile environment. The time was right. We needed our own house.

In the spring, I found a house, which I purchased, with a mortgage of six thousand dollars. We soon moved into it. The new house needed much work and many things in it, so it was necessary to save everything I could save to make the house a home. Though in substantial debt, we were still happy because we were free—that is, we did not have to depend on anyone else. With my earnings, the debt was not crippling; it could easily be paid. Yet my dream of a car would have to wait.

In April 1930, all of us who worked on the idea of a new type of engine for airplanes were struck by misfortune. Our chief engineer and inventor of that engine climbed into a plane with his pilot and flew to York during a blizzard. Their plane crashed into a monticule. The U.S. government, which had a large multimillion dollar order for those engines, recalled its order. Over two thousand people were left without work. It was a severe blow, for me in particular.

The disease of the Great Depression of 1929 had begun to show its symptoms. Many factories and banks closed. There was panic among the people because those who had money in a bank or in stocks lost everything and because there was no work for anyone. I sought no governmental handouts, because they were designated for legitimate citizens.

Detroit, the automobile capital of the world at the time, suffered especially. The large factories in Detroit, most of which employed several thousands of laborers, now employed just a few thousand. Who—unable to pay rent and buy food—would be in the market for an automobile? Thousands of unemployed workers, looking for work, flooded the factories each morning. If a rumor surfaced that a certain factory was hiring, thousands of workers gathered at that factory; the larger the factory, the more people besieged it.

**Picture 42.1. Men during the Great Depression, Looking for Work (Detroit, 1929)**

We used all our savings, and I worried now about food for my family. I was fortunate that for some time the milkman every day delivered milk to us, but when our debt reached about thirty dollars, he stopped coming. I had not paid my mortgage for some six months, but the company from which we bought the house gave us another three months. Yet I was responsible for the interest, but even that was impossible without income. My father-in-law's situation was not much better, so his family moved into our house. Still, that did not much help, as he worked one or two days a week. It was a trying situation, but one that we had to endure.

# Chapter 43

## Head Otaman Calls a Meeting

ON OCTOBER 11, 1930, I RECEIVED A TELEGRAM from the chief otaman Sich, Dr. Stephan Hrynevetsky—elected as hetman, or head otaman, at a 1922 convention of the group Hetman Sich.[86] He said that he would like me to appear in a "very important case" in Chicago. With the telegram, there was a telegraphic transfer with remuneration for the trip.

I went to Chicago the same day. When I arrived at the Sich house, Dr. Hrynevetsky approached me and asked me to go upstairs and into the big hall.

When we entered the big hall on the floor, he handed to me a large package, comprising a long letter, and said, "Read this letter very carefully, because I want you to read it tonight at the current meeting."

"At what meeting?" I said in amazement.

"I convened an immediate meeting of all district otamen."

"For what reason, if you do not mind me asking?"

"In the case of *this* letter. Things in Europe aren't going so well," he added with sadness.

# The Oath

I opened the package, which consisted of some thirty typed pages. It was an account of a lengthy letter by Sich member Vyacheslav Lypinsky, written or dictated by the union "brothers." I ran through a number of pages and said to Dr. Hrynevetsky, who was sitting opposite me, "Mr. Otaman! This is—why this—this is incredible! I can not believe that Lypinsky wrote it!"

"Please read on."

I read further, and as I got deeper in the long letter, I became convinced that this was not the letter of that old Lypinsky whom I knew from his previous writings. It was written by someone who used his name or by a very ill Lypinsky.

In the evening, the district otamen came together. There were sixteen of us. At the bidding of the head otaman, Stephan Hrynevetsky, I started reading aloud the letter. At the end of nearly every sentence, the head otaman interrupted me and offered some comments condemning the actions of our hetman. Others were bothered by the constant interruptions. The atmosphere got tense.

The most irritated otaman was Dr. Ivan Smuk, of the Second District, who was clearly bothered by the ceaseless interruptions of the head otaman. Several other otamen agreed with him. Dr. Hrynevetsky became silent for a while. Yet after I read the second page, Dr. Hrynevetsky again stopped me to offer more condemnatory comments. All others at this juncture were irritated, and they bid Hrynevetsky to remain silent while I finished the letter.

Lypinsky's letter condemned some of the hetman's actions—the most egregious of which was the notion that Hetman Pavlo Skoropadsky sold Carpathian Ukraine to the Magyars[87] for forty thousand *pengs*. The head otaman, irate, objected to this language of "bargaining" for Carpathian Ukraine.

Dr. Smuk interrupted Hrynevetsky. Everyone, heated, began to speak at once in the meeting room. I tried to calm them down, but that did not succeed.

Then came Ivan Denega, the otaman of the Third District. The room was somewhat quiet in anticipation of Denega's words. Otaman Denega began: "Gentlemen, we were taught by Vyacheslav Lypinsky himself, in his more lucid moments, to respect our hetman, even if he errs. So, even if he has done something that is, so to speak, questionable—even though Lypinsky himself ceased to honor him—we ought to remain loyal. We are disciplined and loyal people. Nothing good can come from disunity!"

Otaman Denega's brief speech was received with warm applause.

The head otaman wanted to say something else, but it was agreed that he had, for the nonce, said enough. Everyone knew where he stood.

I asked to speak and was granted permission to do so. "We see, on the basis of this letter," said I, "that Europe, Ukrainians in particular, is in chaos, and it is—it is difficult to grasp how this chaos has begun. Time will tell us, but until then, are we to act chaotically about the chaos? We can not fix any problems if we are not calm."

With these words, someone from Detroit stated irately that I was in agreement with the head otaman and objected to the actions of our hetman. I contradicted that sentiment and asked for permission to finish my thoughts. The hall was silent, and I continued somewhat animatedly.

"When our head otaman fell under the spell of Lypinsky's subversive words, he did not cease then to be our head otaman. But when our head otaman goes on to renounce our ideology, our foundational principles, as does Lypinsky, then we can refuse to obey him and ask our hetman, Pavlo Skoropadsky, to appoint another head otaman by the time of our next congress. But this, fellow otamen, can be done calmly and respectfully—

# The Oath

without screaming and superfluity of words. We are prudent, intelligent people, are we not? Further, let us look at Carpathian Ukraine—the notion of selling it to the Magyars. That is—that is suspicious! Hetman Skoropadsky is not interested in the forceful occupation of Carpathian Ukraine. Moreover, the Magyars [of Hungary] have no reason to buy it. Consequently, the condemnation of our hetman is mere scurrility and can only have been done by people of evil will—sick persons!"

My words received boisterous, abundant applause. Otaman Dr. I. Smoke stood up and said more calmly, "I propose to ban the head otaman from Hetman Sich until the new order from Mr. Hetman Skoropadsky. Does everyone agree?"

"Agreed!" The shout came from all sides of the room.

Dr. Hrynevetsky responded to nothing. He could not. He was universally declared powerless.

Still, he sat in disbelief. Trying to alert others to and to avert a crisis, he had been stripped of power. He could not understand how those otamen who had listened to him and obeyed him unconditionally for fifteen years would now disempower him because of certain "subversive" sentiments. Yet he had the best interest of Ukraine at heart, and he could at least be proud that he had taught his "sergeants" to be straight-line patriots.[88]

"Who will inform Mr. Hetman about our decision?" said Ivan Voitsytsky, the stouthearted otaman from Cleveland.

"Central," replied Dr. Smuk.

"No, no!" said the head otaman. Colonel Alexander Shapolval, as you know—he—he arrives tomorrow night, and he's coming from Europe to see the editor of *Sich*. He can—let him be the informant."

"Fine!" said in almost one voice everyone present.

So went the October 1930 meeting of the otamen and the unexpected withdrawal from authority of the long-standing head otaman, Dr. Stepan Hrynevetsky.

# Chapter 44

## "Like fish on ice"

Picture 44.1. Chemny with
Col. Alexander Shapolval, Years Later

# The Oath

THE NEXT DAY, OCTOBER 12, 1930, was Sunday. After Divine Service, Dr. Hrynevetsky invited me to lunch to speak with him. He was very upset, as he was still in disbelief over being ousted as head otaman. We spoke about many things other than what was really on his mind, and he used sarcasm and wit to hide his true feelings. In his heart, I suspect, he recognized that the sergeants were right to depose him.

At lunch, he asked me to go to Chicago with him to meet with Colonel Shapolval and relate the news. In the early evening and after dinner, we went by car to the railway station.

In Chicago, we met Colonel Shapolval. The three of us went to a restaurant for coffee and a sunny snack. After some banter, Dr. Hrynevetsky, with obvious regret and melancholy, informed the colonel about the course of yesterday's meeting. The colonel, knowing of Hrynevetsky's dedicated leadership over the years, said, "So many years—how—how could *that* have happened?"

"Well, it *did* happened," I said with certain gravity. "You, as colonel to His Highness, have a duty to inform the hetman and to ask for further orders."

"Yes, indeed," acknowledged the colonel.

We spoke about other things for a while. It was already past midnight. I asked Dr. Hrynevetsky to take me to the station so I could go home. He dropped me off at the station, where I waited for the next train to Detroit.

At church on the following Sunday, I was approached by a fellow Sich member, Yakiv Boyko. He told me that the Brotherhood of Michael the Archangel wished to begin a school under the patronage of the parish but away from the church. They wished me to be a teacher at the school and to teach Ukrainian and Ukrainian culture to the children. "The fraternity has allotted fifty dollars per month from its treasury for this purpose," said Mr. Boyko. "The rest will be paid by relatives of children. Are you interested?"

Being out of work and flattered to be recognized for my intelligence, I replied, "Of course! When—when do you think the school will begin?"

"Soon. Here, come to this address on Sunday—two weeks from today." He gave me a piece of paper on which the address was written. "Then—then, we'll have every—that is, all things—all details—ironed out."

"Excellent," I said with ebullience.

We parted.

I left for home in a joyous mood. I told the family about Mr. Boyko's proposal and of my acceptance. My oldest daughter was excited that her father would be a teacher—and a teacher at a new Ukrainian school—but my wife was less than excited.

"Fifty dollars sounds nice, but—but it'll scarcely cover bus fare each day over the course of a term!"

"Well, it is something—better than not working. Besides, I shall be teaching Ukrainian children. I am sure to enjoy that!"

In the afternoon on the assigned day, I arrived at the address. I met Mr. Boyko and found myself surrounded by parents of the children whom I would be teaching.

Mr. Boyko, as representative head of the Brotherhood, led the meeting. I became acquainted with the parents and wrote down the age and degree of education of all the children. Almost all spoke their mother tongue, but four did not know the Cyrillic alphabet. Parents of some seventeen children were present, but I was told that there would be more. Not all the parents could attend the meeting.

Mr. Boyko suggested that parents should pay one dollar per month for each child. Some parents objected that this was too much, especially since some parents wished to send more than one child. After some haggling, it was settled that payment for one child would be one dollar per month; for two children, two dollars per month; for three, two dollars and fifty cents; for four, three dollars. That, of course, would give me a little over twenty dollars per month—scarcely enough on which to live—but Mr. Boyko promised also to draw from the Sich treasury to cover the rest of the promised fifty dollars.

I would be teaching the children three times a week—Tuesday, Thursday, and Saturday—and each day for two hours.

The school opened immediately—two days after our meeting. Having given all the children a preliminary "placement" exam, I divided them into four groups because some of them who went to high school already had some courses in Ukrainian studies.

Instruction went well. The children were obedient and willing to learn. Only one, Ivan Trituruk, was a bit of a miscreant. He was an only child and somewhat spoiled on account of it.

The children made good progress. Their parents and I were pleased. Yet trouble, as the saying goes, never sleeps. In the spring of 1931, in view of the Depression, the state of Michigan issued a decree that all non-locals should be registered by the state because the state wanted to be cleansed of illegal immigrants. That order caused me considerable panic because I would be on the list of those who were to be returned to their native land. It was thought overall to be cheaper to return home illegal immigrants than to house and feed them here. So, early in May, I told my students that we would finish our school year by the end of May because I was likely moving to Ohio.

Instruction was over at the end of May. On a specified day after classes, Father Mykola Strutynsky and I met with the parents. I spoke well on behalf of the children to their parents, and exceptional achievements were duly noted. Father Strutynsky then orally examined some of the children. Parents then questioned the students and me. The children answered well all the questions from the parents, and the parents were happy. The priest then also spoke on behalf of the students. Children thanked me for my instruction. Some brought me monetary gifts in envelopes. One family even gave me five gold dollars. It greatly pleased me that everyone appreciated my instruction, and though I was appreciative of the money I received—I greatly needed it—it meant more to me that my instruction was given a positive moral assessment.

I left the school and was readying to leave Michigan to avoid extradition. Fate intervened. The Supreme Court decided that Michigan's extradition order was unlawful.

Without a job, I searched everywhere for work. Friends in Cleveland found for me a job at an insurance company. In July 1931, we moved to Cleveland, where I became an insurance agent.

This job promised me twenty-five dollars a week, but it was part of my job to find one client each day. I immediately wrote all my friends, but I did not have so many. Moreover, because of the poor economy, people each day were dropping their coverage and there were few others who were adding coverage. The company gave no incentives for the number of people one contacted—effort was of no consequence—but based remuneration solely on new clients. They also charged each agent a fee for any client who missed a payment. So, it was impossible for most agents to earn a living. Instead of the twenty-five dollars pledged to each of us, most of us earned seven or eight dollars per week, an amount on which a person could barely survive and on which one with a family could not. To make matters worse, my wife became ill, and I had not money to call a doctor.

Like a fish on ice that was looking for a return to water, I looked everywhere for work but found nothing. My father-in-law wrote me and requested that I send my wife and children back to Detroit so that they could live with him while I searched for work. Yet I had insufficient money for a train or bus trip.

In September, an acquaintance of mine intended to drive from Cleveland to Detroit. Explaining my inconvenient, but temporary, circumstances, I asked if he could take my family with him. With tears in our eyes, we parted, but I did not grumble about the bitterness of fate. I asked the man to drive carefully and to bring my family safely to Detroit. He promised to drive cautiously.

They left. Alone in the house in which I was staying, I cried. With tear-filled eyes, I dropped to my knees in front of an image of Christ the Savior and prayed like I had seldom before prayed. Why was I being tested so? Why could I not keep a simple, well-paying job and keep my family with me? I had never asked for much from life, but it seemed that the little for which I asked would never be allotted to me.

My neighbor, the elderly Mrs. Volynsky, in whose house I lived, was knocking at my door for a long time. She had wanted to invite me to lunch with her. I did not open the door. I could not open the door. I lacked the courage to appear to her in such a despondent state.

Late in the evening, I went to a shabby little diner to get a cup of coffee and a piece of bread.

## Michael Chemny

My wife and I had decided that we would not remove our furniture—what little we owned was nice—to Detroit. We would instead try to sell everything in Cleveland and live off that money for some time. I managed to sell the furniture, but even then I experienced hardships. The buyers could not pay the full rate at once but pledged to give me a small amount once per week or once per month. I sent to my family the money I got for the furniture. I lived on the few dollars I got as an insurance agent.

In a few days, I received a letter from my wife. She informed me that they had had a horrible accident on the way to Detroit, and they were fortunate that nobody was seriously injured. The driver, trying to pass a slow-moving truck, drove too far to his left, ran into a ditch, and then hit a telegraph pole, which broke into two parts. The dangling part of the pole crashed into the roof and removed a large part of it and broke every window. By chance or the hand of God, my oldest daughter, who was sitting in front with the man as the car rambled through the ditch, happened to turn around to face her mother at the time of crisis. Had she not turned, shards of broken glass would have disfigured her face. The motor was still functional and so they managed the trip. It took time for me to glean all the details of that accident. I thanked God that they were alive—alive and well.

In December 1931, my apartment was barren, and I was completely broke. I left my miserable job, where I had to pay for my other people's insurance premiums. I collected my belongings and left for Detroit, where I settled down with my wife's relatives. All of us were in tears when I arrived, but I was together that evening with my family. That was the only thing that mattered.

The Christmas holiday season arrived. It was not a joyous occasion. It was a custom of ours to give each person some new article of clothing for Christmas, but I had no money for any new clothes. There was scarcely money for bread.

# The Oath

Christmas Eve was bittersweet. On that day, the good women from the Hetman Sich and the Sisterhood of the Immaculate Conception of the Blessed Virgin Mary brought us a large basket of all kinds of foods and canned goods so that we also could have a blessed Christmas. When I saw the basket, I burst into tears. It was impossible to restrain myself. Being a proud man, I was not in the habit of receiving gifts from anybody, but I wanted my family to have something that Christmas, and so I gratefully accepted their kind gift.

Through my tears, I could manage only, "Thank you!" I then rushed out of the room so that the women would not see my flood of tears. Perhaps my behavior seemed rude to them. I do not know. They could not know what was in my heart, so bitterly sad but also so grateful that there were others so kind. I could not help but recall my mother in Khom'yakivka, enjoining me, when still a young boy, to take bread, cheese, and butter to the houses of the most needy of my fellow villagers on pre-feast days.

During the Christmas holiday season, I looked everywhere for work. The men with whom I spoke nodded their heads in understanding. They said nothing because they could say nothing that I wished to hear. I thus passed the winter, spring, and summer.

# Chapter 45

## "Business or Politics?"

DR. OMELYAN TARNAVSKY, in the autumn of 1932, was appointed Otaman Sich. I was instructed to go on a trip around America to revitalize our cells and to establish new cells where there were no cells. I proudly accepted this offer because I hoped that during my trip I could find work of some sort for myself when meeting new people. My hopes were soon dashed. Things were just as bad in all other American cities as they were in Detroit.

I made reports in Cleveland, Akron, Pittsburgh, Philadelphia, New York City, and Buffalo. I made the most improvement in Philadelphia, where I managed to revitalize the First District. There was setting up the services as otaman of a very energetic and loyal person—Mr. Vasyl Volchansky.

My trip did revitalize the Sich movement. Sich Central in Chicago was very pleased with my work. The otaman of the Second District, Dr. Ivan Smuk, suggested at the meeting of the Central Committee that I be awarded a gold medal for my dedicated work. I demurred. "The best medal for me will be the growth and development of our cells."

I remained in Chicago at Sich headquarters for two weeks. During that time, I wrote "My Impressions on a Trip to America" and two short dramas, "Mother and Strilets" and "For Freedom," which were printed in the *Sich* in February and March 1933.

In early March of the same year, I got a sorrowful message from my Khom'yakivka: my dear father had passed at the age of sixty. That distressing news tore at my soul. Had I not left for America, he would likely have still been alive. Over thirty years old, I felt like an orphan in a foreign country.

My brother told me in a letter that my father had a great funeral. Not one but three priests presided over the funereal ceremonies. Not only did everyone from our village attend, but also many people from neighboring villages took part in the funeral. That gladdened me. I thought of my maternal grandfather's great funeral. Both were large men, and large men deserve large funerals.[89] The loss of my father was not something that I would ever get over.

Things would get blacker. I remained unemployed and penniless. I could not even give my poor children a glass of milk each day. I was then not too proud to beg on the streets for pennies so that I could at least feed my children. I did not care about myself, but about my family. There were weeks, not days, that I survived on next to nothing. Every morning and evening, I anxiously prayed to the Lord God for help and strength to endure this crisis.

In August of 1933, I got work at a Ukrainian milk factory as a milkman. The work paid little, but it enabled me to buy bread and to have milk in the house. We lived very economically, and behaving thriftily, we could buy furniture for the house in which we had now lived for six years.

In February 1936, I was appointed as otaman of the Fourth District of Hetman Sich to replace Pylyp Demian. Work in the district required a lot of time for me because I had to set up the affairs of the flight school for our airplanes.[90] Particularly burning was the issue of flight instructors. We appealed to the secretary of internal affairs in Washington, D.C., and they sent us two instructors who taught the theory and practice of flying. Yet maintaining the school and training potential pilots for the airplanes required money. The problems redoubled when Ukrainian pilot and instructor Antin Tchaikovsky fell ill and died. He was buried with great military honor.

**Picture 45.1. The Airplane *Ukraina* (August 19, 1934)**

We were eventually forced to hire strangers, non-Ukrainians, as flight instructors. That proved problematic, as they were unconcerned with the development of the flight school and had no investment in the mission of Sich. In spite of such obstacles, the flight school took steps forward and was a feather in the cap not only of the Fourth District but also of the entire Ukrainian community in metropolitan Detroit. There were several field exercises with the airplanes *Ukraine* and *Lviv*, and many people came to our shows. Such performances, or "flight celebrations," earned us some money—many who attended gave donations for our cause—for school and airplane maintenance. Yet our pilot school was salt in the eyes of our enemies—especially the Ukrainian communists. Still, I must add that many of our trained pilots enlisted in the American Air Force and fought bravely in World War II and were awarded for heroism.

The Moscow embassy in Washington, D.C.—at the behest of a certain Troyanovski, a Pole—requested that the Department of State investigate our schools, and so we were "visited" by agents of the Federal Bureau of Investigation. Those agents, having fully examined our patriotic mission and having noted our generous involvement in American political affairs, not only left us alone but also lauded us for our patriotism and involvement.

I had my enemies on account of my work in the district and undivided commitment to Sich. Thus, there were sent to the supervisor of the Ukrainian milk factory several anonymous letters complaining that I was more interested in politics than in delivering milk. The supervisor showed me some of the letters but mostly ignored them. When the letters proved ineffectual, two men went to the supervisor and told him that if he did not fire me, they would no longer buy milk from his company.

The supervisor had had enough. He said to me, "Business or politics?"

"There's human malice here—hatred, jealousy! It is—it is not about—it is not a matter of business or politics." I paused. "Well, if I am really bad for business, can you move me to a different neighborhood—a different route—where—where there are no—no or few jealous, malicious people? You can fire me, too," I added with a hint of resignation.

"No, I can't fire you, because—well—because you're generally a good man—a good man and a dedicated worker—and, well, I want to keep you as long as possible. But—but you need to be more cautious. Just deliver milk, and keep politics out of your work!"[91]

"Thank you!"

I was given a route in the western part of the city. That convinced me that my personal political enemies had influenced my supervisor. When I asked him why I was given a different route, my supervisor replied that the man who had the route had quit and left for Canada. While that was true, he did not explain to me why I and no one else was chosen to replace him.[92]

Because my position threatened my livelihood, I asked Central Hetman Sich, which had been renamed the Union of the Hetmanites-Statists,[93] to find another district otaman. Central Hetman Sich grudgingly accepted my resignation. They provisionally appointed as otaman Stepan Gava, a Ukrainian military officer.

While I was still otaman of the Fourth District, engineer Petro Zaporozhets came to Detroit. Zaporozhets had previously held the office of secretary-general at Central but moved to Detroit because of a better job. Within a few months, the sergeants of the Fourth District named, on the proposal of a certain Mr. Gava, Petro Zaporozhets as district otaman. Central approved of the nomination, and Zaporozhets became the new chief otaman.

**Picture 45.2. Chemny at Work on the Roof of His House on Ryan Road**

His appointment was auspicious. It occurred on a fateful and historic mission in the autumn of 1937 to celebrate in Detroit the successor to the hetman's ceremonial baton (булава), the hetmanich,[94] Danylo Skoropadsky, son of Pavlo Skoropadsky. The ceremony was breathtaking and majestic. It was beautifully described in the book issued for that occasion—*For Ukraine.*

Having freed myself from involvement as otaman of District Four, I could put more energy into my job. As a result, I earned a bit more money—enough to enable me to purchase an old but mechanically sound car, which I really needed for my work. The effects of the Great Depression were lessening, but many thousands of people in Detroit were still out of work.

Saving money by living frugally, I bought, off Ryan Road and just outside of Detroit, onehalf acre of land, which had a small garden and a small brick building that could be used temporarily as a residence. The building, without water and having no sewage system, needed considerable work, but I was never a stranger to work. I worked by day, and at night I worked in and around the house to make it livable. This double shifting took a toll on me, but I was happy to have a little parcel of earth to call mine. The purchase put me in debt, but I had work, and a little debt was not the worst problem for me.

# PART VI
# TAKING ROOT IN AMERICA

# Chapter 46

## First Ukrainian Congress

I HAD ENOUGH ABOUT which to worry, so I decided to do something about becoming a citizen of the United States. Not having any documents, I applied to my native village to send me a certificate of affiliation. Nonetheless, villages at that time did not yet have the right to issue such certificates, and my request went to the headman of the Region State. In a few weeks, I got from the regional chief, Yuristovski, the following letter in Polish.[95] Because the application was made by my lawyer in Windsor, Dr. Ivan Jatsiv, the answer came to him, as if I were living in Windsor.

*Regional Chairman of Stanislavivka*
*C/o Headman Yuristovski*
*Public Relations Manager*
*Mecislav Rappe, Head of Division*
*Ch. OA 7/24*

*24 April 1937*

*Mykhailo Chemny*
*218 Victoria Avenue*
*Windsor, Ontario*

*On the basis of Article 1, Item 1 of the Decree of the Council of State*

## The Oath

*Defense from 8 November 1920, Dz. V.P.P., Ch. 81, Pos. 540, the Polish State Government denies Mr. Mykhailo Chemny—the son of Kornylo Chemny and Anna Hryniuk, born on 30 November 1901 in Khom'yakivka and to the Khom'yakivkan community—affiliated citizenship in the Polish government because he voluntarily left service in the Polish Army and is abroad Poland and evading his fulfillment of the formal obligation of military service.*

*Therefore, the Regional Government reports that the deprivation of Polish citizenship does not exempt Chemny from all other legal consequences stipulated by obligatory requirements for abandonment of the duty of military service.*

*From this decision, the right to file an appeal to the Ministry of Internal Affairs through the local Regional Government in the period of fourteen days after the day of serving is used, with which the appeal can be brought directly to the authority that issued the decision or through the General Consulates of Poland in Ottawa. The fourteen-day term will be recognized when, before the expiration of the appeal, the appeal will reach the Polish Postal Government, or telegraphic, it will be guided through the consulate. Together with you, your wife, Theophila [sic] Tymchyshyn, is denied Polish citizenship.*

Having obtained this letter as a response to my request, I asked the lawyer to stop the case. I myself would answer that letter and use the same address. I wrote in Ukrainian the following letter.

*Mykhailo Chemny*
*218 Victoria Avenue*
*Windsor, Ontario*

*25 April 1937*

# Michael Chemny

*To Regional Chief Yuristovski of the Office of Stanislavivka:*

*Having received your letter Ch. OA. 7/24 of 14 April 1937, I was very surprised about your decision concerning me, a modest person. You, as regional chief, probably forgot that I am a Ukrainian, not a Pole, and your decision strips me of something I never had—Polish citizenship. I never felt like a citizen of Poland because I always considered Poles to be unwanted intruders in our native land—assailants without any historical or moral rights. I did not ask you for a certificate of affiliation to Poland, but a certificate of affiliation to my community where I was born and grew up without Polish intrusion. If you believe you have the right to grant or exclude my belonging to my native village, I can say only that that decision is idiocy. Any sensible person understands that. You must understand that. For me, a Ukrainian who loves his family and country and who is willing to shed his blood for her liberty, it is not an honor but a dishonor to be a citizen of the Polish state. I would be ashamed and disgusted to be a Pole. I shall not appeal to your meaningless decision. That would be a waste of paper. Still, I am grateful that you have answered my request, though you need not bother yourself further with this matter.*

*Yours,*

*Mykhailo Chemny*

In a few weeks, I received a letter from my youngest brother. As a result of my letter to Yuristovski, the Polish police had vandalized my late father's house, destroyed family mementos and all the photos that I sent from America, and treated my mother very violently. Those were the people who called themselves Christians and considered themselves to be intelligent people. One might expect such behavior from a wild Mongolian. Thus, it was no disgrace to be denied citizenship from such a barbaric state. And so, I put in abeyance my petition concerning citizenship of the

United States and waited for more favorable circumstances.

I performed my duties well as a milkman and showed my supervisor that I had a refined eye for business. So, I was offered a position in the office, where I was responsible for collecting unpaid debts. My route was given to the very jealous man who had all along wanted it.

Working in the office, I could see that the business was headed for bankruptcy. It was in substantial debt, and the customers who owed the milk factory money were not paying what they owed. And so, there were days that the employees did not get their pay on time. I thought it wise not to wait for the business to close. Therefore, whenever possible, I began to look for another job.

I soon[96] told my supervisor that I was quitting because I could not see any future. My resignation was received with regret. The other workers, upon discovering that the factory was soon to close, reproached me for not telling them about the catastrophe. I replied merely that I foresaw no such catastrophe. I did find a better job. Within a few months, the milk factory closed its doors.

On May 21, 1937, I got a job at Snyder Tool and Engineering Company, which produced automated machines and tools. That was the best job that I ever had. Not only was the salary good, but the work also was engaging, and I much enjoyed the work. There I worked dutifully and hard, and soon I became an instructor in the engine department, where I worked until retirement.

Having an ongoing and reliable job, I devoted myself to public-organizational work. I took part in the organization of the Ukrainian Congress Committee of America (UCCA), which worked on offering financial and educational assistance to Ukrainians, and the Ukrainian American Relief Committee (URC). The UCCA and all other of our national superstructures had many obstacles to overcome because Ukrainian socialists wanted everything to go through their hands. And so, there was a congress in Cleveland where not much was decided and progress of the UCCA was stalled. There were numerous long and tedious meetings where we had to persuade people that the UCCA should be one strong organization, not two, so that we could consolidate our efforts and spend money on helping Ukrainians in need, not on debate about helping Ukrainians in need, which was all we seemed to be doing.

Working in those organizations, I became acquainted with many prominent Ukrainians, with whom, in the name of unity and solidarity, I found a common language, which was the creation of one Congress Committee. There were the founder of URC, Dr. L. Tsegelsky, whom I had already known for a long time, and new acquaintances, like Dr. M. Chubaty, Dr. O. Granovsky, D. Halychyn, Dr. V. Galan, and Dr. L. Mishuga, as well as the lawyers A. Shumeyko, M. Diznak, P. E. Rogach, and many others. Other significant Ukrainians—such as Colonel O. Shapolval, B. Katamaja, P. Batyuk, V. Dovghan, I. Kuzya, S. Yarema, and I. Danchuk—I knew from meetings and work related to Sich.

The group was so politically diverse that it was difficult for us to be in agreement on any issues other than the need for unity and the aim of helping Ukrainian immigrants. We did manage to call the Ukrainian Congress to the city of Philadelphia in January 1941.

My fellow Ukrainians dutifully responded to the call, and the First Congress of 1941 went unexpectedly well. Yet the socialist-minded Ukrainian Workers' Union blocked progress and threatened to withdraw from the congress. The result of the congress was formation of four subsidiary organizations: the Ukrainian People's Union, the Ukrainian Catholic Society "Providence," the Ukrainian Workers' Union, and Ukrainian National Aid.

In spite of tense debate, the congress was successful. It was evident that the love for Ukraine and Ukrainian values took precedence over the varying ideals of each congressman.

At the First Ukrainian Congress in Philadelphia—following the report of Dr. L. Tsegelsky, who made his case very clearly—it was decided to create a Ukrainian committee of assistance, the URC, to help our people fleeing from the insanity of communism, many of whom were homeless and without human assistance. Dr. V. Galan, a humble but highly intelligent person from Western Europe, was elected chairman of the committee. He would work diligently but anonymously, as he wanted no recognition for his labors.

There was a problem. Before that, in the Midwest, a similar committee was headed by Ivan Danchuk. Soon there was a dispute concerning who, in fact, was empowered to carry out the duties of the URC. Galan had been elected by the First Ukrainian Congress; Danchuk had been working for two years in the Midwest and had been working chiefly on moral and material assistance to Transcarpathian Ukraine.

And so, we formed a committee to decide the issue of empowerment of the URC. This committee was provisionally headed by Ivan Kuz, with me being the secretary. Later, it was headed by Ivan Danchuk.

After long meetings, it was decided that the committee in the Midwest was subordinate to the committee elected by the First Ukrainian Congress. We then added "United" to Ukrainian Relief Committee; hence its name was changed to UURC.

At first, UURC worked modestly, but upon receiving a congressional mandate, it extended its activities so that it could be of service to thousands of our Ukrainians.

I had always been a part of the committee for the Midwest, but I asked that the committee have the approval of the Ukrainian Congress. Only the congress could validate the Midwest UURC. Congress gave its approval, and we became the United Ukrainian American Relief Commit-tee (UUARC), and in April 1941, we organized a public fundraiser to help our poor Ukrainians and took in several thousand dollars. I was the secretary of that committee and later the chairman of the Controlling Commission. The treasurer was Mr. Ivan Zablotsky, who faithfully per-formed his duties. We coped well with our mission, and UUARC gained great financial support from us.

I suggested that UCCA and UURC create their own affiliates or departments in all the places where we had members. Thus, we could do a more thorough job of collecting money and taking care of new Ukrainians in the United States.

The UCCA was initially indifferent to my proposal, but the UUARC was not. Yet soon the UUARC also accepted my proposal and created divisions in places that helped to realize the goal of helping Ukrainians in need. Though both the UCCA and the UUARC adopted my proposal, neither organization gave me credit for my useful innovation. That did not matter to me. I was just glad I did my part by having a useful suggestion and that both groups were helping Ukrainians. All of us live once, and each of us deserves to be free. Some have to fight to earn that freedom. I was glad to be of some service to my fellow Ukrainians.

I took an active part in all the congresses whenever I could. At the Third Congress, held in Washington, D.C., I was appointed as secretary of the congress. Even then, I noticed bitter partisan antagonism. Perhaps that was why I was elected secretary of the congress. Many others wanted the position because of its prestige. Uninterested in the prestige, I was happy merely to do the dirty work.

## Chapter 47

# "You feel a Holy Duty, and you act on it!"

IN THE BEGINNING OF 1943, I Registered with the government and paid my taxes. During registration, I asked the examiner whether or not I would be facing deportation because of registration.

"No," he said. "Not unless I can find Poland on the map. *Then* we'll deport you!" he added with a friendly chuckle.

On August 9, I got civil papers without the help of the Polish "warlord." I was no longer an illegal alien. Two years later, my wife and daughter did the same. After twenty years of worrying about deportation, we were finally free and happy. We finally got the right to live in the land of liberty—the land of Washington.

It is hard to put into words the perpetual anxiety one feels when any drunkard, any scamp, poses a threat to your life. The slightest incaution through something said or written can be an act that leads to deportation. There were numerous snollygosters among Ukrainians whom I dared not to encounter face to face, for their crafty specialty was to create discord among our people while doing Cain's work. Thus, it was difficult to express openly my hostility to the factionalism among Ukrainians. And so, it was difficult to tell others that Ukrainian strength was not in disunity but in Christian benevolence that led to unity of focus and vision. Only a free man could speak freely and plainly, as only the heart of a free man was unburdened by the notion that he could say whatever he wanted to

## The Oath

say without fear of reprisal. Yet until I was legally free, I could never be sure that my family and I were free of danger.

There were bright moments—though they were few—and only such moments allowed me to embolden my sorrowful soul so that I could continue to fight against inhumanity and evil. The brightest of those moments was a visit to my modest home by His Reverend Constantine Bohachevsky on July 4, 1943.

How honored was I that he should pick me to visit! However, it infuriated many others, even parishioners, who, for haughty, wicked reasons, watched surreptitiously while they muttered, "Who is *he* that the bishop should come to his house?" I ignored those jealous persons. Why should their scurrility surprise anyone?

And so, I hosted the bishop and five Ukrainian priests. We had several toasts to Ukraine prior to the sumptuous dinner, prepared by my wife and her sisters. After dinner, we walked among the cherry trees in my garden, and the bishop told me that he chose to visit me out of gratitude for my unwavering patriotism and defense of spiritual authority through word and pen. "I shall always remember what you and the true Hetmen have done for the cause. On account of such actions, we have been able to better conditions in our diocese. Please tell your brothers that I am grateful to you—all of you!"

"Thank you very much for the recognition, Your Excellency! It is our holy duty!"

**Picture 47.1. Bishop Constantine Bohachevsky with Priests, Summer, 1943**

"Yes, yes! But you feel a holy duty, and you *act* on it! How many others . . . ."

We conversed about different things for a while, and then the conversation turned to Father S. Tichansky, pastor of the Church of the Immaculate Conception of the Blessed Virgin Mary in Hamtramck, who was in the yard with us. When done, I brought in a large handful of cherries.

The bishop visited for three hours and departed in a beautiful limousine, driven by Mr. Andriy Hradovsky. The bishop's visit made my wife, children, and entire family feel blessed, and my friends began even more to appreciate and respect me.

The singular event was filmed by my friend T. Lytvyn, but it was lost at some point after the death of his wife.

# Chapter 48

## World War II and the Demise of the UHO

WORLD WAR II, WHICH HAD begun in 1939, affected American commerce. There was no free exchange of goods, and Americans began to worry about the fate of the "free world." Thus, America became an ally of Moscow for the purpose of defeating Hitler. The Muscovites took care that the liberties of the people in the Russian lands had been reduced to a minimum.

Our UCCA, even the United Hetman Organization, in which I took an active part, became an object of investigation by American authorities because many Ukrainian refugees found passage to America through Germany.[97] Agents from the Federal Bureau of Investigation appeared in the districts and demanded to see our books—especially books on Ukrainian protocol. Our center, the Ukrainian Club, was also raided be-cause of suspicion that Ukrainians were cooperating with the Germans and rejoicing in their military accomplishments.

The chief concern was that our United Hetman Organization's parliamentary organization was pro-German. Some of our people who worked in Central were afraid of that insinuation. Our lawyer, B. Delyhovich, predicted that our group would self-destruct. Our cells were ordered to give all their weapons to American military authorities, although those weapons, purchased with our own money, were mostly obsolete. Our printing house was sold for nothing. The SHD (СГД) House was simply handed over to the UNC department in Chicago without any remuneration. So our noble organization, which had a beautiful and long history, ceased to exist in 1942.

As there seemed to be no hope for the future, we continued to live for the past. The UCCA continued to meet. We collected money to assist helpless Ukrainians, and we rescued many from starvation. We produced affidavits for the refugees—we found ways to do that without the assistance of Germany—and helped them to come to the United States. In Detroit, we gave assistance to 127 families.

In November 1948—after the war—we gathered and decided to reorganize the Hetman group. And so, an Organizing Council consisting of twelve people was created. Of the twelve, there was a five-person executive committee: the chairman, the secretary, the treasurer, and two others. The executive was supposed to revise the charter of the organization, create a name, and incorporate it.

Petro Zaporozhets was elected as the chairman of the executive board; I, Mykhailo Chemny, as secretary; and Mykola Nebozhenko, Stepan Gava, and Yevgen Draginda as the other executive members. I was the principle author of the mission statement, which Zaporozhets translated into English. The organization's name, United Hetman Organization (UHO), was suggested by Hetmanich Danylo Skoropadsky. Eight people were on the board. In Detroit, two divisions were initially organized, and later, a third. At the First Congress in America, which already had ten cells in America (UHOA), a new administration was elected in keeping with the statute, and the Organizing Committee and its Executive Committee were now defunct. As of this writing (c. 1964), the organization has fifteen departments spread across the United States.

# Chapter 49

## Two Life-Threatening Surgeries

I WAS TORMENTED BY an ulcer as early as World War I, and that ulcer got worse, increasingly painful, as I got older. I saw several doctors for my condition, but no one was of any help. The pain over time had gotten so severe that I reconciled myself to imminent death.

On April 27, 1949, I went to Hamilton, Canada, where I had a friend, Dr. Yaroslav Berezovsky, who was recommended to me. He took X-rays of my stomach, and those showed that I needed to undergo surgery. I agreed to the surgery.

When I went home, I told my family about my condition and about the surgery, slated for May 8. My children and wife objected that surgery was too risky—they were afraid of losing me—and I, in order not to make them sad, met with the doctor on May 7 and canceled the surgery. Yet the doctors assured me that if I wanted to live, I would have to undergo an operation in the autumn at the latest. Further delays could mean death.

Once home, it was back to work, though the pain was insufferable. On October 3, while at work, blood came from my stomach and shot out through my mouth. A doctor was called, and I was taken to the hospital, where I stayed for several days in readiness for surgery.

I went to the operating table on October 8. The surgery lasted three and three-quarters hours, and I had all but one-eighth of my stomach removed.

The surgery embittered me. How could I live with such a small stomach? My wife and children sat at my bed in the hospital to keep me in good spirits and to make sure that I remained alive. I prayed to the Lord that he might allow me to get up from the hospital bed and to have some semblance of a regular life in my remaining years. My wife and daughters also prayed.

**Picture 49.1. Chemny and Wife (center) Flanked by (left to right) Daughters Yvestafiya and Natalka (1939)**

Our prayers were heard. In three weeks, I left the hospital and went home. Yet the doctor told me that in six weeks, when I was stronger, I would have to undergo another surgery because I had an inflammation on my side and he was worried that the lump was cancerous. On December 6, after having received the Holy Sacraments, I was subjected to a second operation. The inflammation, it turned out, had diminished because the source of the inflammation, an ulcerated stomach, was gone. Thus, it was a common, benign inflammation, and there were no signs of cancer. I was relieved.

The operation much weakened me, and I had been waiting for a while until the wound healed. Yet the doctor prematurely removed the stitches, and the wound was reopened. With the wound reopened, it became infected. I could not take in any food and was given a medicine that greatly weakened me.

I was thus confined to bed for three months. At the beginning of the fourth month, February 1950, I got myself out of bed. Not long thereafter I returned to work at Snyder's, where my boss and fellow workers were glad to see me back.

# Chapter 50

## Trip to California

WHEN I WAS HOSPITALIZED in October 1949, I was elected as vice-president of Providence—the Union of Ukrainian Catholics. For this honor, I was grateful to the members of Providence, but unfortunately, I was not healthy enough to take part in the inaugural meeting. Later, already at their biannual meeting, I took an oath and participated in the business affairs of that noble organization.

The work in Providence was not new to me. For many years, I had worked as the head of their department in Detroit, #162. Nonetheless, the new position imposed on me new, larger responsibilities. I had to attend different conventions, meet with people on behalf of the organization, investigate various cases in the departments, and be present at the biyearly meetings in Philadelphia, among other duties. Yet I took things slowly, and the Lord blessed me with enough health that I coped well with my tasks for four years.

After my first term, I volunteered for a second, although I did not take any measures to be reelected, as did the others candidates who were looking to be elected vice-president. My plate was already overfilled. Shortly after being elected, another would take my place.[98] I was not worried about that because I, in addition, had a lot of work in Motherland, Bulava, the UCCA, the Native School (Рідній Школі),[99] and the UHOA, where I was always picked for high posts. At the same time, I had to work as the founder and four-year secretary in the Ukrainian Self-reliance Cred-it Union in Warren, which had grown from almost nothing to a large five-million-dollar institution. Yet most of the time, I was involved in the

# The Oath

UHOA and in the Ukrainian American Club, where I had been manager and secretary of the Council of Trustees for years.

In July 1953, I went to California to visit my cousin (тіточні сестри) Olga Miller and, maybe, to find a better life. I had heard much about California but had never been west of Chicago. The trip by rail was pleasant, and passing through ten states, I traveled through diverse, beautiful landscapes that changed like the multicolored patterns in a kaleidoscope. In Kansas, the golden wheat stretched to all parts of the horizon. It reminded me of Ukraine. In Colorado, the majestic peaks of the Rockies, some still covered with snow, touched the heavens, while the valleys were deep, lush, and verdant. Arizona, a virtual desert with red and well-worn mountains and numerous wind-slapped boulders, also had a singular charm. Last, Southern California, with its fields of grapes and citrus, had a remarkable beauty that does not exist in the North.

Upon my arrival in California, my cousin was excited to see me. We had not seen each other in twenty years. Enjoying my time with her, I stayed in California for several days—longer than I had expected—and I saw as much of California as I could see. After mass on Sunday, I asked about the whereabouts of my old friend Prof. Vasyl Yemetz. We did not stay in touch through letters, and all that I knew was that he lived somewhere in Hollywood.

I called him from my cousin's house, and he rejoiced that I was so close and invited me to his home. We agreed to meet the next day, a Monday, in the evening. On Monday, I took a taxi to the much-ballyhooed Hollywood. My dear friend the professor and his wife were waiting excitedly for me. As there was a long path from the road to his house, they came halfway to meet me. I was warmly and courteously greeted. I did not know Mrs. Yemetz because my friend got married after our acquaintance. She was kind, handsome, and very witty, and she could speak well both English and Ukrainian.

We entered their living room. I was seated. Mrs. Yemetz appeared with a bottle of good wine and some snacks. When the wine was finished, she invited me to dinner. I declined because I had already been to dinner at my cousin's.

We sat for a long time and conversed on a variety of topics of significance—especially the Hetman's movement. Prof. Yemetz had fallen in love with the movement as a boy and served faithfully all of his life the movement's ideals, whether through his music or essays in newspapers or journals. Though known for his music, he had in recent years devoted more time to the literary field, where he published many fine essays on public issues. His famous essay, "To the New Poltava," had a large readership among our people, many of whom condemned the essay. They said Yemetz, under the pseudonym O. Voynarenko, made old sins—sins that created today's socialists and justified slandering conservatives. His unpardonable sin, said they, was the overthrow of the Ukrainian Hetman state, which led to Russian yoke, the exile and imprisonment of our brothers and sisters, and the death through starvation of twenty-five million Ukrainians.

# The Oath

**Picture 50.1. Vasyl Yemetz,
Virtuoso of Ukraine**

Prof. Yemetz, however, had no direct connections with events in Ukraine,[100] although he subscribed to a large number of Ukrainian news-papers and magazines, and so he merely accepted at face value the news I shared with him about the actions of the Hetman's movement and, in general, about Ukrainian life.

We sat and talked for hours. Around midnight, the good Mrs. Yemetz ordered me a taxi and I returned to my cousin's home. Since my visit, the professor and I have become close friends. We continue to correspond to this day.

# Chapter 51

## The Hetmanich Pays Me a Visit

Picture 51.1. Hetmanich Toasts Ukraine
(Chemny's House, Outer Drive, October 23, 1953)

ABSORBING MYSELF IN THE work of the above-mentioned institutions, the days swiftly passed. Yet the work was stimulating and important—it needed to be done—so I was unaware that I was getting older.

# The Oath

My daughters married, and each now had her own life to live. Though my wife and I were happy, we felt that our house was now too big for us, so we sold our house on Ryan Road with its large garden and moved to a smaller house on Outer Drive in the city so that my wife would be close to her mother and younger sister, Gloria. At the new house, I quickly began the work to make the house not only a home but a Ukrainian home (оселю українця).

On October 23, 1953—it was a Friday—I had another singular, unforgettable experience. That day, late in the evening, I had the great pleasure of welcoming the second most noteworthy guest in my home—His Highness (Його Світлістю) Hetmanich Danylo Skoropadsky.

We met the hetmanich at the Willow Run Airport that night, and after exchanging greetings, we went by two cars to my house. As it was already late night, we toasted, had a small snack, and retired to bed.

On Saturday, UHOA held a meeting at the Ukrainian Center, over which Hetmanich Skoropadsky presided.

Near evening and after the meeting, we went back to my house for dinner. At the well-stocked table prepared by Mrs. Eugene Draginda and my wife, there sat a distinguished group of patriots: Hetmanich Danylo Skoropadsky; Dr. Myroslav Siminovich, head of the department of higher education; Petro Zaporozhets, the deputy of the hetman; fellow members of the Brotherhood; Michael Pervak; Eugene Draginda; Prof. Michael Ovchinnik; Michael G. Boyar; the teacher Mykola Pasika; Juliani Zebrynsky; Rosalie Draginda; and my wife and me. There was serious political conversation, but there was also levity. All took pictures of that significant event.

On Sunday, His Highness, my wife, and I went to a Ukrainian Catholic church on the west side of Detroit. After the divine service, the hetmanich traveled with a large escort to the Ukrainian Orthodox Church of the Protection of the Blessed Virgin Mary, where, on the occasion of the special holiday, there was Bishop Mstislav.

After the liturgy, we went to the Ukrainian School Hall, where there was to be a great celebration organized by the Ukrainian youth and the Women's Department of the UHOA. The youth cordially congratulated the hetmanich. In response, the hetmanich gave a very meaningful statement, urging them to unite and endure in our struggle to a free Ukrainian state.

Dinner and many photos left us with indelibly etched memories of that historic event.

After this reception, there was a short meeting of the Hetmen, over which His Highness presided. The meeting much motivated all of us to spread the Hetman's movement on American soil.

At eleven p.m., we drove our hetmanich to the airport, where he was in the company of teacher Yevgeniya Ziblikevich, who flew with him to New York. We cordially bid farewell to His Highness and, with elevated spirits, returned home. None of us would know that this would be the last time we had the pleasure of seeing that proud, courageous face and listening to his wise counsel. He would be killed in three years and four months.

# Chapter 52

## Another Trip to California

**Picture 52.1. Second Visit to California**
Chemny, my cousin Olga Miller, Theophilia Chemny,
and Rosalie Draginda (1955).

# The Oath

IN JULY 1955, I ONCE again headed to California—this time, in my car with my wife and with the company of Eugene and Rosalie Draginda. We intended to visit many different places, even Mexico.

One of the first activities was a trip to the shore with Prof. Yemetz, his wife, Stanley Kostiv (my brother-in-law), and his wife to enjoy the beach and swim in the Pacific Ocean. We went to the famous Malibu Beach.

It puzzles me how the Pacific Ocean got its name, because it is anything but pacific. Even on a mild day, when the winds are slight, the Pacific Ocean always throws up tall waves that ravage the shores and warn beach-dwellers, "Swim at your own risk!" Though the day was nice and the winds were slight, we could not swim. The sun was trying to poke its way through thick fog, which I learned was present every morning, but it was not too warm. We reclined on blankets thrown onto the sand. It was delightful, relaxing. At one point, I decided to try a swim, but the water was too cold, and I returned to the beach. And so, we relaxed on the beach for some two hours and dined with what food the Lord had blessed us and spoke on various topics.

"Cover your legs at least with a towel, brother, or you'll get burned," Prof. Yemetz said sincerely.

"But—but, there's no sun! How can I get burned?" I answered.

"You do not know the Californian sun," he said with a laugh. "Though you can't see it through the fog, it'll burn you!"

"How strange." I covered my legs, but it was too late. When we left the beach in a few hours, my left leg was hot, as it had gotten too much sun. Prof. Yemetz suggested we go to his home, a new house nestled in the Hollywood mountains. So high was the house that it stood in the sky at an equal height to City Hall of Los Angeles.

The good Mrs. Yemetz took us in and treated us as a kind hostess, but my leg had really begun to hurt me. Mrs. Yemetz was much concerned about my condition. I wanted to go to the pharmacy and get some lotion to reduce the pain. By evening, my leg, which had been scorched under the sun's rays, was nearly beet red, and I felt such a burning pain—a pain of such intensity that I had never before felt—that I could not sleep. Mrs. Kostiv, with whom we drove to the Yemetzes' household, advised me to climb into a tub of hot water to alleviate my pain. Though I thought little of her "womanish remedies" (бабські ліки), I gave it a try. Not only did that not help, but it also severely worsened my pain. My left leg was swollen from top to foot, and the pain was unbearable.

We had to prepare to head back to Detroit because our vacation time was coming to an end. Early in the morning, we left Altadena, California. I was driving, but my aching leg pain made driving next to impossible. However, I managed by evening to make it to Flagstaff, Arizona, where we spent the night. I was then in a fever. To fight the fever, I went to the city and asked for a good apothecary and took from him an ointment to alleviate the pain. Still, I was in great pain for almost the entire night, and so I could not fall asleep till morning, when it was nearing the time to wake and resume the trip back home.

We were on the road for three more days, and because of my painful leg, much of the driving was done by my friend Yevgen Draginda. When we arrived in Detroit, my leg was better and I could once again freely walk around.

I shall never forget that trip to California for two reasons. First, I had a welcome and warm visit with my cousin and her family. Second, I suffered through a pain the likes of which I had never before felt.

# Chapter 53

## Death of Danilo Skoropadsky

Picture 53.1. Gravestone of Chemny's Parents (Khom'yakivka)

# The Oath

FEBRUARY 23, 1957, WAS A VERY sad day—not only for the United Hetman Organization, but also for all patriotic Ukrainians. On this unforgettable day, we heard from London the devastatingly sad message that our dear hetmanich had died.

The loss was unfathomable; many were inconsolable. Ukraine's biggest hope for the future had passed, and with his death, hope for Ukrainians also passed. Without our leader, there was great dissention among members of the UHOA. It was difficult to convince patriots, even those with the best intentions for the motherland, to think as one when the unifying force no longer existed. The All-Hetman Congress in Detroit in July 1958 aimed to normalize those relations.

After the congress in Detroit, all of us recognized that there was much work to be done in the organization. With the passing of Hetmanich Skoropadsky, a new superstructure in the UHOA was created. It was now necessary to work with double enthusiasm, for there was double the work to do. Thank God, the work followed both in content and in form a smooth course.

I worked somewhat unobtrusively in the different organizations of which I was a member, until another fateful day, December 24, 1962. At Holy Supper at my home, I felt great pain in my chest, and my doctor could not be reached. I had to wait for him until Christmas. He told me that I had suffered a light heart attack, and I was taken, via ambulance, to the hospital, where I spent three weeks before returning home. That day marked a steady decline in my health. On May 18, 1963, I again fell ill and wound up in the hospital. In ten days, I returned to work.

In early 1965, I began to have severe joint pains, and my legs became weak. I had many doctors examine me, but none have been able to help.

That is where I am today, as I write these final sentences. My debilities not only do not leave me, but also they worsen. Whenever I walk a few dozen steps, I get out of breath and my painfilled legs falter so that I have to sit or lie down. How long this will continue, only the holy saints know! It seems that I, too, am ready for my grave![101]

As I reflect back on this journey, my life, I have had many broken dreams and I have suffered much. Still, I have had a good life. I have been true to myself, to my family, to God, and to my Ukraine. Ah, my Ukraine! I reflect back to that oath I took on November 3, 1918, and I realize that my sufferings and broken dreams were not pointless. I have always been true to my Ukraine. If I could go back just once more before dying and see again my father and mother, then, well . . . .[102]

Michael Chemny

# EPILOGUE

Those "tired" words with which Chemny ended his book were not prophetic. He would live for another fourteen years, though they were somewhat unfruitful years due to a steady decline of his health.

It is certain that much happened in those years, as much happens in a fourteen-year span of any person's life. That Chemny wrote the book when he did and ended it when he did likely speaks to the notion that he really did think that he had little time left.

It is unfortunate that so little is known by me and his other grandchildren of those years. His daughters have passed, and so there is no way of gleaning information from them. I enter what I have gleaned from my own recollections and from my brothers David and Edward and from cousins and fellow grandchildren Stephen Fedak and Andrea Nowak; but unfortunately, much has been forgotten, and of what can be remembered, precious details are absent and must be filled in by reasonable inference.

We do know that he retired from Snyder Tool and Engineering Company in 1966 and was given a portable television, with engraving, as a gift for retirement. Chemny always dreamed of becoming an engineer in America, and recall he spent two years at the Polish engineering institute in Detroit, but reality always got in the way of his schooling in that regard. Snyder's gave him an opportunity to exercise his creativity and intellect in ways other than in simple manufacture.

Early in the 1970s—after the passing of my father, Edward Holowchak, on October 19, 1970—Chemny, his wife, and his wife's sister Anna, who lived with them at their Outer Drive residence, moved to Sioux Street in Redford Township, Michigan, to be nearer to my mother, Natalie, who lived in Redford Township, so that they could be of help to her, if needed.

Chemny's health continued to falter. His heart was a constant problem. My grandfather had two heart attacks, including the one in 1962 he mentions, and two debilitating strokes. Though he continued reading, his intellect was no longer so sharp. While living at Sioux and driving his Cadillac on Inkster Road one afternoon, he became confused and pressed the accelerator instead of his brake, and his 1966 Cadillac crashed into the front porch of one of the houses on the north side of the street. He recovered but would never again drive, though he would often talk of getting himself a "little Pinto" or "little Nova," as if driving a much smaller car would make him less of a liability on the road. Not being able to drive was devastating to him—a blow to his manhood and personhood and a signal that what was left of his vitality was gone, or nearly so.

The strokes, both of which occurred after 1970, were incapacitating. The first stroke left him with partial paralysis in his left arm and left leg. The second was crippling and led to severe loss of mental functioning. He would die a few days after the second stroke, on February 15, 1979—eight months prior to his eightieth birthday.

Short in stature, Chemny was stout in character. He was intelligent, imaginative, loyal, and courageous. He loved his Ukraine, and though he would never again return to it once he got on the boat to cross the Atlantic, he would serve Ukraine as one of its most faithful citizens.

Yet the measure of a man—any man—is, in large measure, the memories others have of him and the influence he has had on others. Chemny had five grandchildren, and each has memories of him—some large, some trite. And so, I thought it apposite here for each of his five grandchildren to share one short story.

The oldest, Stephen Fedak, writes: "My fondest memories of my grandparents include traveling to their house on holidays and on the weekends for family meals that my grandmother made for us." Yet one special moment concerning Chemny sticks out. "Being the military buff that he was, he made sure that he and my grandmother attended my commissioning ceremony to Second Lieutenant in the U.S. Army in 1970. I was so happy, and proud, that he was there to witness the commissioning along with my parents, sister, and my soon-to-be-wife, Mary Kay."

Granddaughter Andrea "Panda" Nowak remembers his generosity. When she was in need of her first car and could not be helped by either of her parents, Chemny volunteered his time to be with his granddaughter and help her in her crisis. He ushered her to some four used-car dealerships and helped her pick out her first automobile—a Ford Maverick. "He was always there for us, even if he had things to do," she sums.

My older brother David remembers fondly working and playing with Chemny. "I spent many long hours working with and for him when he couldn't do the work for himself. I learned a lot from him, and I now treasure those moments." He recalls also that Chemny, with a bit of prodding, would enjoy playing the board game Monopoly with us. "I remember how Mark and I used to try and get him to play Monopoly with us. He would always refuse at first, but our pleas would inevitably wear him down. Then Mark and I would make sure that he got his favorite color group and that he built houses and hotels on it. It was more fun watching him win and collecting rent from us than us winning. Every time we paid him, he would smile and say, 'Happy to do business with you.'"

My story is comical. On November 12, 1972, seven years after his tired words to close *Trampled Dreams*, Michael and Theophilia Chemny celebrated their fiftieth wedding anniversary. It was a gala event. My grandfather and his wife left Immaculate Conception Ukrainian Catholic Church and readied to drive to the VFW Hall at Conant and Moran in Hamtramck. I had driven down with them. As we walked to his darkgreen 1966 Cadillac, my grandfather, perhaps tongue-in-cheek, said to me, "Marky, you sit with me in the front seat. Fifty years with one woman—that's too *goddam* much!" My grandmother smiled and willingly took the rear seat, as I opened the door for her. The two were fêted at the hall by family and their numerous friends.

My younger brother, Edward—whose memories are less vivid, given he is the youngest of the grandchildren—recalls a sad scene in his room at Providence Hospital after Chemny's second stroke, early in February 1979. Chemny lay in bed with a dumbfounded look. A nurse bent forward near his bed, and he touched her blond hair with the curiosity of a child. Thereafter, he said nothing. It was a sad moment for Edward, who recognized that Chemny was no longer Chemny.

**Fiftieth Wedding Anniversary.** In November, 1972, Chemny and his wife, Theophilia (seated third and fourth, bottom), family, and friends, celebrated 50 years of marriage.

All of us, his grandchildren, remember one constant. Whenever we would come to visit him at his Outer Drive residence or even at his residence at Sioux, when he was still relatively robust, he would, weather permitting, be seated outside, in his garage, or on the driveway, with a magazine, newspaper, or book in his hands and an unfiltered cigarette in his cigarette holder. He was a voracious reader throughout his life, and he knew much about many things. He learned, from his youthful days in Khom'yakivka, that it was impossible to get ahead in life without knowledge, hence his preoccupation with the Ukrainian Reading Room. One might even say that he recognized that the atrocities he had witnessed in the Old Country were due to human ignorance.

It is now fitting to say a few words about my grandmother, Theophilia Chemny. That she was a seldom-mentioned, behind-the-scenes figure in this "autobiography" does not mean that Chemny had little affection for her or that she was to him a relatively insignificant figure in his life. She was not. She was his bedrock. I know. I lived with my grandmother after his passing, for seven years.

Though a quiet, unassuming figure, she was a strong, resilient, and dutiful woman who shared her husband's appetence for hard work—hers, of the domestic sort. His many triumphs in life—and there were many—happened because she was always there for him. She disallowed failure, or at least made difficult failure, by doing the "little things," without which the "large things" could not be done, and doing them without regard for fanfare. For instance, once established in Detroit, Chemny often had large "political" dinners at his residence at Ryan Road or at Outer Drive (recall the visits of the bishop and the hetmanich); and Theophilia, sometimes with the aid of her sisters, would prepare those dinners, serve their guests food and drink, and clean up after the dinners. All those things were done noiselessly (revisit Picture 51.1), as if the male guests were at some exclusive restaurant, where they were to be doted on.

Still, it must be acknowledged that it was probably not a relationship of mutual love—at least, not of mutual love of any intense sort. Recall that

Chemny's mother arranged the marriage, and Theophilia did not have any choice in the matter. Her father had consented to the marriage, and that was that. My grandmother was a simple peasant woman who understood little of political matters, and Chemny was robustly political, so in that regard, they were far from compatible. Yet she obeyed her father, married Chemny, and attended to his needs as wife for fifty-seven years—all that, and she did not complain. That is what I remember most of those seven years I spent with my grandmother. She went about her daily domestic chores—and they were considerable—and she never complained. More than any other person I have known, she was indomitable, unbreakable. Thus, she ensured his successes and enabled him to endure often-extreme hardships.

Finally, something must be said about Chemny's title, *Trampled Dreams*, which in certain respects seems to me, having trudged through this book many times, to be a misnomer. As readers will have by now discovered, Chemny lived a full, adventure-filled life. Thus, one wonders why he chose the title that he did.

Chemny's life was, for himself and for his fellow Ukrainians, a quest for freedom—a freedom to live as each would choose, not as others would dictate. That, he never achieved. The world of his day was too raw, and the people of his day were too mercurial—too readily moved by the shifting winds of events on account of greed and ignorance.

Yet throughout his life, Chemny had the purest freedom that anyone could have: freedom of conscience, or authenticity. Chemny's actions were always consistent with his words, and that sort of freedom demands profound courage that too few in his day, and in our day, possess. That sort of freedom is the same whether one lives in Khom'yakivka or in Detroit, whether one is a milkman or an engineer, and that makes his story, more than others', worth remembering.

## Michael Chemny

There was one constant in Chemny's life—love of Ukraine. He never faltered in that love, and when he removed to America, he was always a Ukrainian in America. As a Ukrainian in America, he did what he could from afar to work toward Ukrainian independence, and he lived in Michigan consistently with those Ukrainian values and customs that he so dearly loved. Thus, he was always true to his oath and had more triumphs than he had trampled dreams, hence my preference for a change of title from *Trampled Dreams* to *The Oath*.

# ENDNOTES

[1] This can also mean "village farmers." I use throughout the less pejorative term "villagers."

[2] The text says north, and that is correct if we consider that parts of the winding river lie south of Budzyn, but Budzyn lies immediately southeast of the Dnister River.

[3] This is the only mention of Fr. Verhanovsky in the book. We are not told when Fr. Michael Durdello takes on Verhanovsky's duties in Khom'yakiv-ka.

[4] The translation mentions a bench upon which the two sat and a place "inside" with fragrant hay, where Grandpa would lay down his grandson. It is possible that they would go to a specific place in the woods, perhaps an abandoned farm.

[5] "Perhaps ... grandpa" added to the text.

[6] Chemny uses селяни, which gets translated as "peasants," and that is a possible translation, but "farmers" or "village farmers" is the most common meaning and is, of course, less pejorative.

[7] This paragraph I have added to my grandfather's text.

[8] Another source indicates that children would wait for the brightest star to appear—a signal of Jesus's birth. This source also excellently confirms Chemny's account of the meal. Suburban Grandma, "Ukrainian Christmas Eve Tradition," Culture, 20 Dec. 2016, http://suburbangrandma.com/culture/ukrainian-christmas-eve-tradition/, accessed 9 July 2017.

⁹ This passing comment, along with several references to honey, show that many farms had apiaries, perhaps merely the wealthier ones. Beekeeping is today one of the foremost economic activities in Ukraine and it is one of the world's largest producers of honey.

¹⁰ One must assume that Christmas mass would begin at four a.m., the time at which mass for Yordan would begin.

¹¹ Chemny uses the term Myasnytsy. Maslenitsya might be the oldest Ukrainian holiday. It has its origin in Slavic mythology as a festival in honor of Volos, the sun god. Thus, it was originally a festival in honor of the beginning of spring. For Christian Ukrainians, the festival occurs during the week prior to Lent, so it is a time for gaiety.

¹² A Ukrainian alcoholic beverage, derived from hority (burning) and similar to American moonshine. Horilka is sometimes used generically to mean vodka or any strong alcoholic beverage. Chemny makes it clear in chapter 10 that horilka means Ukrainian vodka.

¹³ "… or at least mostly" was added to the text. Chemny typically exaggerates the orderliness of such celebrations. It is hard to believe that, with everyone drinking vodka, there was never raucousness.

¹⁴ Ukrainians are noted for their exquisitely ornate Easter eggs.

¹⁵ Sometimes called "Passion Gospels" because they recounted the Passion of Christ, these were traditionally read on Holy Thursday.

¹⁶ Ukrainian Catholics make the sign of the cross in the manner of Orthodox Christians. With thumb and first two fingers together in recognition of the Trinity, they touch their forehead and then lower stomach to cover the vertical part of the cross, and then they touch their right shoulder and then their left to cover the horizontal part of the cross.

17 This text I relegate to a footnote: "World War I brought an abrupt end to this tradition of dressing girls for Easter and sharing clothes. Girls became resentful and insincere. They began to turn a blind eye to tradition. They became attached to their old clothes and refused to share them."

18 The descent of the Holy Spirit on the Apostles on the fiftieth day after Christ's resurrection.

19 This account was originally placed after my chapter 11, but it is clearly out of place there.

20 Where the angel Gabriel, sent by God, announced to Mary that she would be the mother of Jesus.

21 In this and the previous paragraph, Chemny begins with the young ones leaving their houses for the fields at night to sing. He immedi-ately turns to the young men reaping the crops. He ends with returning to town with dawn nearing. I assume that all this efficient reaping was done at night, hence I add "by the light of the moon" in the prior paragraph, which is not in the text.

22 Chemny's text reads, "... though never more than one tiny glass."

23 "Each took off ... begin mowing" occurs after the description of the two large fields in Chemny's text.

24 The Ukrainian says the former was 150–160 morgs; the latter, 110–120 morgs. I was unable to find any definition of "morg." Yet Chem-ny tells us Krichman was "nine square kilometers long," which I interpret as "nine kilometers long and wide," but that might be an exaggeration of size. If the interpretation is appropriate, then Krichman is 81 square kilometers or roughly 31 square miles, and Voonyava is 110 square kilo-meters or roughly 42 square miles. Each is an extraordinarily large field.

²⁵ The first translation says, "It was mowed later, because the hay got dry earlier there." I remove the casual claim because that would seem to be a good reason to mow it first.

²⁶ Held in early August in remembrance of Christ, after rising from the dead and on a mountain, showing his disciples his divinity.

²⁷ This was originally chapter 2 of Chemny's text, which jumps from Chemny's life at ten to his life at fifteen. It chronologically fits bet-ter here, as the next chapter turns to World War I, which Chemny de-scribes from September 1914.

²⁸ I added this to the text.

²⁹ The year had to be 1914, when the war started. Chemny tells of deeds occurring in the winter of 1914–1915. Thus, September 14 was a Saturday.

³⁰ This sentence does not appear in Chemny's account, but it is clear later that he, too, was in Tysmenytsya to watch personally the pil-laging.

³¹ Chemny does not say whether the Cossacks were Russian or Ukrainian, but it is clear from later text that they must have been Rus-sians or—if Ukrainians—Ukrainians fighting in the Russian army.

³² Chemny's original text reads, "He then threw a silver chalice into the crowd, which, of course, would always form whenever there was any noise in the village." This seems untoward, as the presence of the Cos-sacks is not just any "noise."

³³ I originally took this as boldfaced irony, but soon changed my mind, as future references to the priest show Chemny's great regard for him. Chemny consistently writes of Fr. Durdello as a kind and caring

# The Oath

man of God.

³⁴ Choice of words here shows a very young Chemny expressing Christ-like wisdom, and that must be taken with a grain of salt.

³⁵ The translation reads, "... then finally stopped in the alder-tree forest." It is unlikely that such a raging fire would stop in the forest, hence my preference for "before."

³⁶ The text reads, "... became sincere and good people for some time"

³⁷ I surmise here, as the text is illegible.

³⁸ This sentence originally appears at the end of the second paragraph below: "My mother ... went to sleep."

³⁹ The bit "and too little to drink" added by me. It seems obvious that someone engaged in trying to stop the spread of a wildfire all day would not stop to take in all the fluid he needs for such an arduous task. Moreover, dehydration is a large cause of exhaustion.

⁴⁰ The translation reads, "Our people cannot understand what they are said to [understand], they want to see the getting better of their destinies and lives...." The sentiment, I suspect, that Chemny aims at is ease of living and material gain.

⁴¹ Here another translator began and finished the text, as the first died prior to finishing the task.

⁴² The original text has also a "breeder," which I have omitted.

⁴³ This paragraph appears at the end of Chemny's chapter 14. I have removed it to the beginning of this, the following, chapter, where it better fits.

44 The translation reads, "Maybe as much as bad will not be comforted I, but in my heart I rejoiced that at last will fall apart a prison of nations."

45 The translation just mentions "large," and that could be applicable to the square instead of the First Battery.

46 The claim "and I care not to write more about that" I added.

47 The translation has "Whipping the horses."

48 This paragraph added to the text.

49 For more, see "Central Rada," Internet Encyclopedia of Ukraine, http://www.encyclopediaofukraine.com/display.asp?linkpath=pages%5CC%5CE%5CCentralRada.htm.

50 Some details here added to Chemny's text.

51 Again, much was added to Chemny's sketch.

52 For more on Pavlo Skoropadsky, see "Skoropadsky, Pavlo," Internet Encyclopedia of Ukraine, http://www.encyclopediaofukraine.com/display.asp?linkpath=pages%5CS%5CK%CSkoropadskyPavlo.htm, accessed 6 Jan. 2018.

53 The translation read, "where has a reasonable power."

54 The translation has "Klybivtsi," not "Klubovets," but Chemny makes no mention of Klybivtsi below.

55 The translation reads, "invited us in...." That is likely not a translational slip. Chemny was probably part of a small group of young men, of which he was the lead, entrusted with the significant task.

56 The translation reads "him," but Chemny above relates that his horse is female.

57 Translation reads "filming."

58 A small house.

59 The translation has "two waterings," which I have excised.

60 Literally meaning "enlightenment."

61 The translation reads, "With songs on our lips, we went on a path that was fading, sparkling with its rare swamp against the flames from the already-gone sunshine."

62 It might be more reasonable that Chemny shattered bones in his foot as a result of the grenade. It seems strange that a grenade would effect a clean break.

63 The translation, jumbled, reads, "'If this does not make you much difficulty, then I'll arrive in three days,' *** said. 'You see, typhus is thrown between the shooters, and I would like to even send those who are not too sick to somewhere in the fresh air, so that they, besides their wounds, did not have to fight with typhus.' I cradled along the crests with the hands of comforting and comforting them, spending long days and nights."

64 This last sentence I have added.

65 An open space for celebratory events or festivals.

66 This sentence I added to make a reasonable transition. The translation reads, "The doctor examined patients, prescribed medicine, gave us instructions on how to deal with them and asked him to take him to Stanislavivka. I was tired of traveling, so sent my brother, giving him

money for medicine, and reminded me not to be very hurt by the horses, because they have already done today in a 30 kilometers. The brother went to the horses, adjusted the seats for the doctor and they departed. I was left with the sick. My sister Efrosina sincerely helped me, although she was only 14 years old. We made cold compresses, fed sick relatives, and often there were days that none of us fell asleep at least one minute a day." I broke this into two paragraphs and took obvious liberties with the text in some effort to make sense of it.

[67] The translation reads, "we." Did Fr. Durdello go with Chemny?

[68] Original translation reads, "All the hungry will climb in a bowl and a spoon."

[69] A Pole of the Mazurian ethnic group.

[70] Symon Petliura had become the leader of the Ukrainian Direc-torate in February 1919 and signed an alliance with Poland in Warsaw, in April 1920, which conceded Galicia to Poland in exchange for military support in the continuing struggle with the Bolsheviks.

[71] Probably a military officer of some inferior rank.

[72] The tenor of this exchange strongly suggests that there were some men in attendance.

[73] This was originally a lengthy chapter titled "The Cooperative," which I have broken into three chapters, given three different and prominent themes.

[74] These two paragraphs were originally one, and appended at the end of the prior chapter. They fit better here.

[75] One thousand acres.

[76] Originally titled "My Family" for reasons that are unclear.

[77] The metaphor is senseless, as the prior sentence says the light shines on the sea, not on the shore.

[78] "Cassios" in the text. Cassia is water with cane sugar.

[79] The text reads "her," which I have changed to "his."

[80] I have been unable to locate Vyborg in Florida, so it is impossible to know where Chemny, Demyan, and Fritz landed. The southern and western tips of Florida are swampland.

[81] On the fifth day, a Saturday morning, a taxi arrived at Don Juan's Farm. A man with a suitcase got out

[82] Pavlo Skoropadsky (1873–1945) was a great soldier and mili-tary leader. He became hetman of Ukraine on April 29, 1918, by ousting the Rada and its Ukrainian People's Republic. He was ousted from power by an uprising by social-democrat Symon Petliura later in the same year and fled to Germany.

[83] For more, see "Otaman" in the Encyclopedia of Ukraine (http://www.encyclopediaofukraine.com/display.asp?linkpath=pages%5CO%5CT%5COtaman.htm).

[84] The choice of words seems strained, as Chemny is a staunch supporter of the status quo, which involved a hetman, in effect a Ukrainian king with unlimited powers, and a hierarchy of religious authorities. There was nothing democratic in his vision of a thriving Ukraine.

[85] A strange sentiment, given the debilitating events Chemny relates in this chapter.

[86] Hetman Sich was a militant conservative Ukrainian American

group that formed a "liberation army" in the hopes of freeing Ukraine. It was common for Sich members to join the American Militia, now the National Guard, as did Chemny, to gain military training for the liberation of Ukraine. In the 1930s, the group purchased three airplanes—Ukraina, Lviv, and Kyiv—in anticipation of military activities. See "United Hetman Organization," http://www.encyclopediaofukraine.com/display.asp?linkpath=pages%5CU%5CN%5CUnitedHetmanOrganization.htm, accessed 5 Jan. 2018.

[87] Hungarians

[88] It is unlikely that Hrynevetsky was ousted for this single episode—disagreeing with the hetman. It is probable that dissatisfaction with his leadership had been building up over time. Consider merely his paternalistic and incessant interruptions at the emergency meeting.

[89] These last two sentences added to the text.

[90] See footnote 88.

[91] The last sentence added. It seems fairly clear, in spite of what Chemny writes, that he has been mixing politics with work.

[92] Chemny here does not consider what he writes earlier concerning his request to be given a different route.

[93] The United Hetman Organization. See "United Hetman Organization," Internet Encyclopedia of Ukraine, http://www.encyclopediaofukraine.com/display.asp?linkpath=pages%5CU%5CN%5CUnitedHetmanOrganization.htm, accessed, 6 Jan. 2018.

[94] Crown prince.

[95] This paragraph was originally the last paragraph of the preceding

chapter.

[96] The text reads, "Early in the 1940s, I ... ," but that makes no sense, given his job on May 21, 1937, in the next paragraph.

[97] The phrase "because ... Germany" I have added because of what Chemny says below.

[98] The translation cryptically reads, "Apparently that's why I crashed in a few voices." The context, given the next sentence, suggests he might have been asked to leave, given inadequate performance due to overwork.

[99] I.e., the Ukrainian school.

[100] The translation reads, "... with the outside world."

[101] Здається, що я вже втомився. The sentiment literally may be taken thus, "It seems that I too am tired," but given the picture of his parents' gravestone (Picture 53.1), it should be less literally interpreted.

[102] This paragraph was added to the book.

# Michael Chemny

www.ingramcontent.com/pod-product-compliance
Lightning Source LLC
Chambersburg PA
CBHW030316100526
44592CB00010B/450